Carliana Rouse Favretto
Andrea Souza Castro

Rainwater harvesting for use in vehicle washing

Carliana Rouse Favretto
Andrea Souza Castro

Rainwater harvesting for use in vehicle washing

Case study for the municipality of Pelotas - RS

ScienciaScripts

Imprint

Cover image: www.ingimage.com

This book is a translation from the original published under ISBN 978-3-330-77174-1.

Publisher:
Sciencia Scripts
is a trademark of
Dodo Books Indian Ocean Ltd. and OmniScriptum S.R.L publishing group

120 High Road, East Finchley, London, N2 9ED, United Kingdom
Str. Armeneasca 28/1, office 1, Chisinau MD-2012, Republic of Moldova, Europe
Managing Directors: Ieva Konstantinova, Victoria Ursu
info@omniscriptum.com

Printed at: see last page
ISBN: 978-620-8-63838-2

SUMMARY

ACKNOWLEDGMENTS

I would like to thank my parents, Joel Favretto and Nelsi G. Schmitt, for their presence, guidance and support.

To my siblings, Eder G. Schmitt and Vanessa F. Schmitt, for their advice, friendship and for being my brothers and sisters.

My friends Carla Volz and Evelise Dal Ponte, for their companionship and friendship at all times.

To my friends and colleagues in Pelotas, those I met and learned from on a daily basis throughout my degree, those who took part in the long days of studies and projects, acting as a hub of help, support and friendship during the difficult and fun times of this period.

The company Expresso Embaixador, especially Mr. Ricardo Almeida Cepeda, for the opportunity to carry out this study.

To Professor Maurizio Silveira Quadro for his friendship, knowledge and advice throughout the academic period.

To my advisor, Professor Andréa Souza Castro, for her friendship, patience and guidance in the development of this work.

Professors Diuliana Leandro, Luciara Corrêa and Rita Fraga Damé for their availability and dedication to the profession.

And finally, to those who are no longer present today, but were responsible for many choices and decisions on this journey. Thank you very much.

SUMMARY

FAVRETTO, Carliana Rouse. **Rainwater harvesting for use in vehicle washing: Case study for the municipality of Pelotas - RS.** 2016. 93f. Course Conclusion Work (TCC). Degree in Environmental and Sanitary Engineering. Federal University of Pelotas, Pelotas.

Rainwater harvesting systems are simple alternatives that help to reduce consumption of treated water, reduce the population's public water supply bill and also dampen surface runoff due to the possibility of storing it. These systems are particularly suitable for establishments that have large catchment areas and need large volumes of water to meet the demands of their activities. The aim of this study was to analyze the potential for using water in the municipality of Pelotas - RS, to evaluate the methods for sizing a water storage reservoir, with the aim of using it to wash the vehicles of a local business and to check the economic viability of installing it. Firstly, daily rainfall data for the municipality was collected from the Pelotas Agroclimatological Station (EAP) for a 36-year period (1985 to 2015). The water demand required for the company's activities was verified through site visits and the reservoir sizing methods applied were the Rippl method, consecutive days without rain, Analysis and Simulation and the Brazilian, German, English and Australian Practices. The economic viability of the system was checked by comparing the possibility of installing it before and after. The results showed that the municipality of Pelotas has the potential to install rainwater harvesting systems, due to the occurrence of regular and well-distributed rainfall throughout the year; however, due to the large volume of water required for the company's activities, the system does not always work satisfactorily. The results of the volumes for the reservoir obtained by applying the different methods were discrepant, but when the water balance inside the reservoir was carried out using the Australian Practical Analysis and Simulation method, a better understanding of how it works and the possibility of making a choice was obtained.

We recommend using the largest catchment area simulated (10,544 m^2) and applying a 455 m reservoir[3], representing a saving of 79% in treated water and a reduction in water supply costs of approximately R$17,652.00 per month.

Keywords: Rainwater harvesting, Car washing,

1 INTRODUCTION

Currently, the subject of "water availability" is a headline in many media outlets, which frequently report on the current situation, cases of scarcity, waste and pollution. According to May (2004), population growth combined with the improper use of water has led to scenarios of water degradation. For Cohim et al. (2007), in addition to population growth, the industrialization process and the demand for water in large urban centers are also factors that contribute to this situation.

According to estimates by the International Food Policy Research Institute, it is estimated that by 2050 a total of 4.8 billion people will be in a situation of water stress (SEGALA, 2012). This problem is justified by the factors presented above, which have generated concern and encouraged programs for the sustainable rational use of water, as well as the search for new alternatives and solutions aimed at reducing the shortage of this resource.

In view of this concern, water reuse, seawater desalination and rainwater harvesting are emerging as alternatives to help with the supply of non-potable water.

These rainwater harvesting systems are used to meet non-potable water demands (NOLDE, 2007), since water of reasonable quality is available for various purposes, especially non-potable water, and can be used in various industrial, agricultural, commercial and residential activities.

In the industrial sector, it can be used for cooling, air conditioning, laundry, machine washing, boiler supply, vehicle washing, fire control systems and general cleaning. In homes and commerce in general, it is usually used for sanitary basins, washing vehicles, cleaning and watering gardens. Finally, in the agricultural sector, it is mainly used to irrigate crops (MAY and PRADO, 2004).

Given this alternative source, rainwater contributes both to reducing the volume of treated water consumed and to saving users' public supply bills, as well as helping to cushion surface runoff by attenuating flooding due to the volume of water retained for storage. These conditions are relevant when they have favorable characteristics for implementation, especially in processes that require high water demand and large areas of coverage for collection (MIERZWA et al. 2007).

In addition, various sectors have come to see the use of rainwater as "*marketing*" for institutions, through the introduction of sustainable practices and socio-environmental responsibility (TOMAZ, 2000).

In urban areas, some authors, such as DEVKOTA et al. (2015) and SANTOS and TAVEIRA-PINTO (2013), comment that the application of this system is still limited due to the long payback period, which makes implementation unfeasible in some cases. In this context, it is essential to carry out a quantitative and qualitative assessment for its installation.

Qualitative factors refer to the elements present in the atmosphere, which can affect the quality of the water collected. As far as the quantitative aspect is concerned, it is important to know the volume of water precipitated in the region and the demand for water needed to meet a given purpose, with the aim of setting up a system that guarantees the supply of non-potable water most of the time and that is economically viable (ANNECCHINI, 2005).

1.1 OBJECTIVES

1.1.1 General Objective

The aim of this work is to verify the environmental, technical and economic viability of a rainwater harvesting system for washing the vehicles of a road passenger transport company located in the municipality of Pelotas - RS.

1.1.2 Specific Objectives

- To evaluate the potential for using rainwater in the municipality of Pelotas - RS.
- Comparing and analyzing reservoir sizing methods for water storage.
- Estimate the reduction in drinking water consumption and the savings on supply.

2. LITERATURE REVIEW

2.1 The importance of water

Water is a natural resource of fundamental importance for the survival and development of all species on the planet (MMA, 2005). Due to the hydrological cycle, the volume of water on Planet Earth remains constant, occupying approximately 70% of the surface, making it one of the most abundant natural resources available on the planet (LOBATO, 2005).

Despite having most of its surface covered by water, when it comes to its availability for human consumption, this figure becomes worrying. According to the World Health Organization (WHO, 2006), around 97.5% of available water is salty and only 2.5% is fresh water.

The freshwater portion is divided between glaciers and polar ice caps (68.9%), underground aquifers (29.9%), soil moisture (0.9%) and finally, the surface water portion made up of rivers and lakes (0.3%) (MMA, 2005).

The availability of drinking water is a long-standing problem and the future trend is unsatisfactory. In many cases, this issue can be attributed to its scarcity due to population growth, waste, inadequate geographical distribution, pollution and degradation of water sources, and illegal use, resulting in compromised development and growth for future generations (SAUTCHÙK, 2004). In addition, industrial expansion and climate change, which alters the distribution of rainfall, are also factors that contribute to worsening the problem of scarcity (HAGEMANN, 2009).

Among the various purposes for which water is used around the world, the following stand out: water supply, electricity generation, irrigation, navigation, city cleaning, construction, firefighting, among others. According to the Ministry of the Environment, around 70% of all water consumed goes to agriculture, 22% to the industrial sector and 8% to domestic activities (MMA, 2005).

The United Nations World Report on the Development of Water Resources (2015) indicates that disturbances to ecosystems caused by excessive urbanization, inadequate agricultural practices, deforestation and the untreated discharge of effluents into water bodies are among the factors that threaten the environment's ability to provide ecosystem services in order to sustain life on the planet.

The sharp increase in the world's population, and inevitably the increase in drinking water consumption, is leading to a reduction in the quality and availability of water resources

(MARINOSKI, 2007). Research carried out by the UN shows that the world's population will rise from 6.6 billion to 9.1 billion by 2050. Estimates indicate that if this is true, the increase of 10 billion inhabitants over the next 50 years will lead to water supply problems for around 70% of the population (CHRISTOFIDIS, 2003).

In addition to the increase in population, its poor distribution around the globe is a worrying factor when it comes to water availability (MARINOSKI, 2007). The MMA (2005) states that countries which have an abundance of this natural resource, such as Brazil, are not free from a water deficit, as availability varies from region to region, causing an imbalance between supply and demand.

The most populous regions have the least water available, while the opposite is true: regions with lower population levels have a greater volume of water available. In this sense, this natural resource is not only limited by its quantity, but also by its quality and location (GHISI, 2006; BRANCO, 2010).

Brazil has an area of approximately 8,515,767 km^2 (Anuârio Estatistico, 2014) and approximately 190 million inhabitants (IBGE, 2010), making it the fifth largest country in the world in terms of both land area and population. According to data from the National Water Agency (ANA), the country holds around 13.7% of the world's available surface freshwater, although the availability of this resource is not uniform.

As a country with a large territorial area, it has a diversity of climates, topographical conditions, vegetation, socio-economic and cultural conditions, all of which make water management a difficult task.

According to the data presented by Conjuntura dos recursos hidricos no Brasil (2013), approximately 80% of the country's surface water availability is found in the Amazon basin, which is inhabited by less than 5% of the total population, likewise, only 20% of water resources are available to 95% of the Brazilian population.

Figure 1 lists the water availability, population and land area of the five regions that make up Brazil.

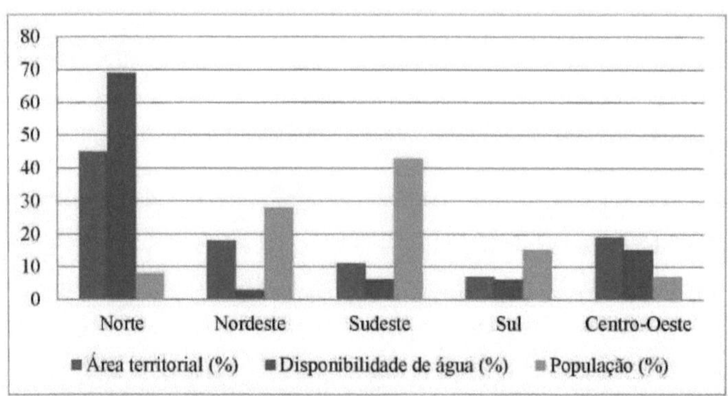

Figure 1. Proportion of land area, water availability and population for the five regions of Brazil. Source: IBGE (2010); ANA (2013).

The imbalance between supply and demand for water in Brazil is evident, especially when you consider that the regions where industrial and agricultural activity is most prevalent, such as the southeast and northeast, have the highest population concentration and the lowest water availability.

In the Northeast, water shortages are chronic. In the Southeast it is abundant, but of poor quality. The invasion of water sources by the poor is one of São Paulo's biggest problems. Industrial waste discharged into the Paraiba do Sul river makes the water that supplies Rio de Janeiro and other cities precarious. There is a lack of water to irrigate rice paddies in Rio Grande do Sul (FREITAS, 2000).

In addition to having a large surface water supply, Brazil has one of the largest groundwater reservoirs in the world: the Guarani Aquifer, which covers an area of almost 1.2 million km^2 and is located in the Parana Geological Sedimentary Basin, between 4 neighboring countries (Brazil, Paraguay, Uruguay and Argentina). Due to its size, the aquifer is the main underground water reserve in South America, with an estimated volume of 46,000 km^3, of which 2/3 of the total area is located in Brazil, between the states of Goias, Mato Grosso, Minas Gerais, São Paulo, Paraná, Santa Catarina and Rio Grande do Sul (AQUIFERO GUARANI, 2015).

2.2 Precipitation characterization

Originally, the word pluvial comes from the Latin word *Pluvium*, which means rain. Thus, the name rainwater or rainwater is a characterization of the water that comes from precipitation.

Precipitation is the release of water from water vapor in the atmosphere onto the Earth's surface, in various forms: dew, drizzle, rain, hail, hailstones or snow, differentiated from each other by the physical state in which the water is found (TUCCI, 2001; VILLIERS, 2002).

Condensation of the water vapor present in the atmosphere is the result of its cooling to the point of saturation, and can occur due to the frontal action of other wind currents, accentuated topography, thermal convention phenomena or a combination of all these causes (GARCEZ and ALVARES, 1988).

When the precipitation event begins, contamination can occur due to the loading of particles that are suspended in the air. These particles are harmful substances such as sulphur dioxide (SO_2) and nitrogen oxides (NO_x). Normally, this contamination occurs in urban areas due to the large number of vehicles and industries (TORDO, 2004).

Rainfall is an important factor in controlling the hydrological cycle and one of the climatic variables that has the greatest influence on the quality of the environment. Climate variability is an important characteristic to be analyzed in the process of installing rainwater harvesting systems, as it will determine the success or failure of their operation.

According to INMET (2016), the state of Rio Grande do Sul has balanced rainfall distributions throughout the year, but the volume of rainfall is uneven in the different regions of the state. The southern region has an average rainfall of between 1299 and 1500 mm, while the north has an average rainfall of between 1500 and 1800 mm.

Several climatic events are responsible for the variation and distribution of rainfall in the state. Known as Mesoscale Convective Complexes (MCCs), these are systems that form during the night and generally show maximum convection over the south of Paraguay and reach the state with intense rainfall in a short space of time (GUEDES, 1985). This event usually occurs in the spring season.

The El Nino Southern Oscillation (ENOS) phenomenon causes significant rainfall anomalies in Rio Grande do Sul, characterized by higher rainfall and thermal changes. In El Nino years, the chances of above-normal rainfall are greater, while negative deviations occur in La Nina years (BRITTO et al., 2008).

A study by Britto et al. (2008) analyzed the state's monthly and seasonal rainfall variability and found four similar sub-regions: northeast, northwest, south-central and coastal.

The south-central and coastal sub-regions had more rainfall during the winter season. In spring, it rains more in the northwestern sub-region of the state, when MCCs are present. In summer, it rains more in the northeastern sub-region and in fall in the southeastern sub-region.

a) Precipitation in Pelotas

The municipality of Pelotas is located in the southern region of the state of Rio Grande do Sul, specifically at latitude 31°46'34" and longitude 52°21'34". The variations in temperature and precipitation in the municipality are associated with air masses and frontal systems coming from the continental and maritime regions.

According to data from the Agroclimatological Station of Pelotas - EAP (2016), which refers to the rainfall station located in the municipality of Capâo do Leâo on the premises of Embrapa Clima Temperado, rainfall can be considered to be well distributed throughout the year, With a monthly average of 120 mm, February is the month with the highest volume of rain, with an average of 163 mm, and October is the month with the lowest volume of rain, with an average of approximately 109 mm (Figure 2).

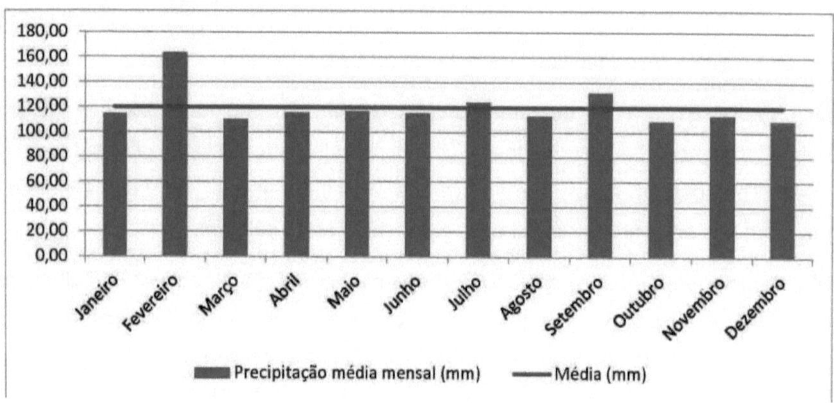

Figure 2 - Average monthly rainfall in the municipality of Pelotas - RS from 1980 to 2015. Source: Pelotas Agroclimatological Station (2016)

2.3 History of rainwater harvesting and examples of systems

Rainwater is an ancient and simple form of water supply. Reports of the storage and use of rainwater have been used by different civilizations over time in various parts of the world. Historical records indicate numerous rock-cut cisterns used for storing rainwater, dating back to 3000 BC (SAUTCHÙK, 2004; MARINOSKI, 2007).

The oldest cistern in the world is probably the one in Istanbul, Turkey, known as the Yerebatan Sarayi, built during the Roman Empire and with a capacity of 80,000 m^3 of rainwater (UNEP, 2002; WERNECK, 2006).

On the African continent, the oldest evidence of rainwater collection and storage is found in northern Egypt, where cisterns have been used for more than 2,000 years, with a capacity of between 200 and 2,000 m3, many of which are still in operation. In China, cisterns and rainwater tanks existed two thousand years ago (GNADLINGER, 2000; UNEP, 2002).

In Mexico, ancient writings indicate the use of this system since the time of the Aztecs and Mayans. There are still cisterns in use today, dating back to before Christopher Columbus arrived in America. In Israel, there is one of the best known examples, the Masada fortress, with ten reservoirs dug into the rock, with a total capacity of 40 million liters (TOMAZ, 2003).

Abanbars, a traditional community rainwater harvesting system, are still found in Ira (GNADLINGER, 2000).

In Germany, the government provides funding for the construction and installation of catchment systems, with the aim of supplying water distribution in the city, saving drinking water and conserving groundwater, which is used as a source of water in many of the country's cities (GROUP RAINDROPS, 2002).

One of the countries that most promotes studies and investments in rainwater harvesting systems is Japan, as well as the construction of detention reservoirs to prevent flooding (TOMAZ, 2003).

In the United States, Australia and Singapore, research is also being carried out in this area. In 1992, a rainwater harvesting system was implemented at Chagi Airport in Singapore. The rain collected on the airport runways is used to flush the toilets (GROUP RAINDROPS, 2002).

In Brazil, the oldest rainwater harvesting system was installed on the island of Fernando de Noronha in 1943 by the Americans (MAY, 2004). Today, the island still uses rainwater.

Collecting and using rainwater is a popular technique, especially in Brazil's semi-arid regions (SOARES et al., 2000). As this region has around 80% of its geographic subsoil area in crystalline formations, with no water table, rainwater storage has proved to be the most suitable option for supplying human consumption (MAY, 2004).

According to the ASA network (Articulaçâo do Semi-àrido Brasileiro), a program called "Programa um milhao de cisternas" (One million cisterns program) was developed with the aim of helping people access drinking water by building plate cisterns and improving their quality of life. Since 2003, approximately 420,000 cisterns have been built in the region (ASA, 2013).

In 1999, the Brazilian Rainwater Management and Harvesting Association was founded in Brazil. It is responsible for disseminating studies and research, bringing together equipment, instruments and services on the subject (ABCMAC, 2008).

2.4 Brazilian legislation and standards for the use of rainwater

Water is a public good, an unlimited natural resource with economic value. In situations of scarcity, the priority use of water resources is human consumption and animal watering (BRASIL, 1997).

Law No. 9.433 of January 8, 1997 created the National Water Resources Management System and, through the management instruments, specifically Article 5, item III mentions the granting of the right to use water resources.

According to CIRRA; FCTH; DTC ENGENHARIA (2004), the grant:

"... gives the managing body the conditions to manage the quantity and quality of these resources, and the user the guarantee of the right to use the water. The granting authority (the Union and the states) must evaluate each request for a grant, checking that the existing quantities are sufficient, taking into account the qualitative and quantitative aspects. In this way, the grant orders and regulates the various uses of water in a hydrographic basin (...).

Generally speaking, rainwater and reuse water do not require a permit because their collection and storage do not directly interfere with the balance of the water body, but they can intervene qualitatively and quantitatively in the project that is adopted.

In the context of sustainable development, the management and administration of water resources can be defined as actions aimed at managing the rational use and conservation of water, guaranteeing compliance with quality standards and controlling the use of water without compromising the environment and the well-being of society.

In this sense, and as a tool for managing water resources, the Commission on Sustainable Development Policies and Agenda 21 comments:

The management of water use and the search for new supply alternatives such as the use of rainwater, the desalination of seawater, the replenishment of groundwater and the reuse of water are part of the context of sustainable development, which proposes the use of natural resources in a balanced way and without prejudice to future generations (BRASIL, 2002).

The planning of rainwater harvesting systems plays a fundamental role in the sustainable management of water resources, providing an alternative in public water supply, savings for users and consequently the conservation of natural resources for more noble purposes.

The Water Code, described by Decree No. 24.643 of July 10, 1934, considers rainwater to be the result of rainfall. It also states that rainwater belongs to the owner of the building where it falls directly, who can dispose of it at will.

Rainwater harvesting is indirectly related to the objectives of this policy, since it encourages rational use and at the same time prevents critical hydrological events, both droughts, due to the promotion of reserves, and floods, due to the reduction in surface runoff. The inclusion of rainwater harvesting in the Plan is indicative of the efforts of water resources policy to achieve cross-cutting and integrated water management (SENRA; BRONZATTO; VENDRUSCOLO, 2007).

In Brazil, there is still no legislation on the sustainable use of rainwater. However, there are several Bills (PL) currently before the Chamber of Deputies, including PL 7.818 of 2014, which establishes a National Policy for the collection, storage and use of rainwater, with the aim of promoting and conserving the rational use of water, environmental quality, proper management and economic incentives for collection, with a view to promoting the direct and planned use of rainwater.

Currently, in federal terms, the use of rainwater is still being shaped by draft legislation, however, in some Brazilian states and municipalities, this issue is already old and demanded, especially in places where water availability is scarce.

In recent years, several Brazilian municipalities have adopted legislation on the use of rainwater, encouraging the development of new habits among the population in terms of rational water consumption and the adoption of new sources of supply. It is believed that through Law 11.445/2007, which makes it compulsory to draw up municipal basic sanitation plans, the number of municipalities adopting measures to use rainwater will increase and spread the importance of conserving water resources.

In the state of Rio Grande do Sul, the municipality of Porto Alegre was a pioneer in adopting measures on the subject. Law No. 10.506 of 2008 instituted the Program for the Conservation, Rational Use and Reuse of Water, with the aim of promoting measures necessary for the conservation, reduction of waste and use of alternative sources for the collection and use of water in buildings, as well as raising awareness among users of its importance to life.

With this in mind, and with the aim of reducing drinking water consumption and maximizing the use of rainwater, technical standards have been developed for the installation conditions of the catchment system.

Initially, in 1989, ABNT published NBR 10.844, called "Instalações prediais de águas pluviais", characterized by the criteria and requirements necessary for the execution of water drainage installation projects, with the aim of guaranteeing levels of functionality, safety, hygiene, comfort, durability and economy, and the standard applies to building roofs, terraces, patios, backyards and so on (BRASIL, ABNT, 1989).

In 2007, NBR 15.527 was published, entitled "Use of roofs in urban areas for non-potable purposes - Requirements", with the following main objective: "To provide requirements for the use of rainwater from roofs in urban areas for non-potable purposes. It applies to non-potable uses in which rainwater can be used after adequate treatment, such as flushing toilets, watering lawns and ornamental plants, washing vehicles, cleaning sidewalks and streets, cleaning patios, water mirrors and industrial uses" (BRASIL, ABNT, 2007).

2.5 Rainwater collection system

Rainwater harvesting systems can be simple or complex, and can contain different types of devices. Simple systems depend fundamentally on three elements: precipitation, horizontal and vertical conduits and the storage reservoir. Complex systems, on the other hand, are suitable for large-scale projects, as they require professional assistance, investments and larger or interconnected reservoirs to store large volumes of water (WATERFALL, 2002).

Regardless of the type of system to be adopted, the local environmental conditions, climate, economic and spatial factors must be analyzed in order to make the system viable.

For Cilento (2009), the rainwater collection surface and the reservoir for storing the water are the main structures that make up the collection system. Tomaz (2003) considers that the most important factors for a well-functioning system are: rainfall at the site, the catchment area and the demand required by the project.

The structures are interconnected via horizontal (gutters) and vertical (pipes) conduits, grids, filters and a storage box for the first rainfall, when necessary. The system can work by gravity or by inserting pumps to convey the water to other reservoirs.

In addition to NBR 15.527 (2007), a partnership between the National Water Agency - ANA, the Federation of Industries of the State of São Paulo - FIESP and the Civil Construction Industry Union of the State of São Paulo - SindusCon (ANA, FIESP and SindusCon-SP, 2005) produced the Manual for the Conservation and Reuse of Water in Buildings, presenting a basic methodology for the design of rainwater collection, treatment and use systems, as shown in Figure 3 and described through the following stages:

- Determination of average local rainfall (mm.month^{-1});
- Determining the collection area;
- Determining the flow coefficient;
- Design of complementary systems (grills, filters, piping, etc.);
- Disposal tank design;

- Choosing the necessary treatment system;
- Cistern design;
- Characterization of rainwater quality;
- Identification of water uses (demand and quality).

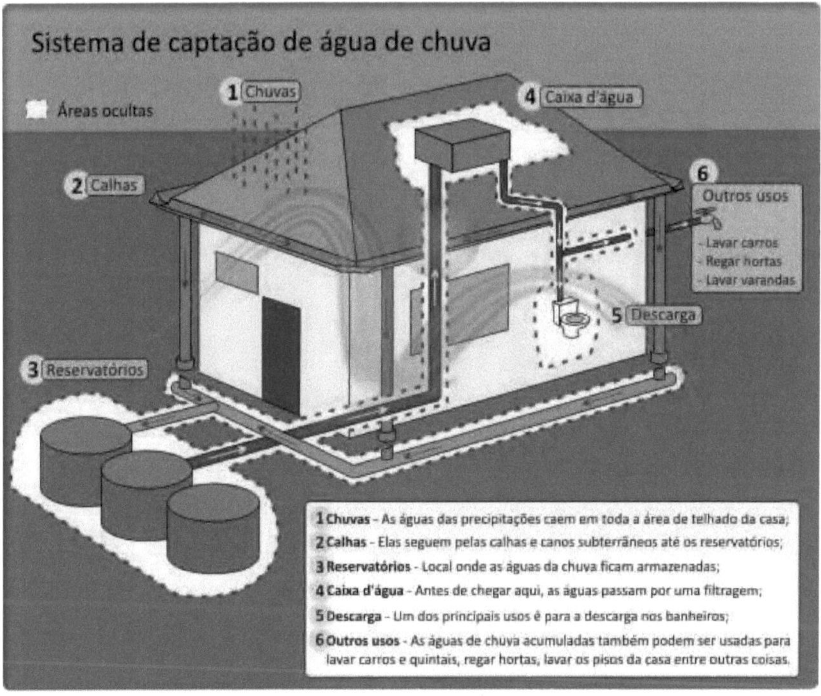

Figure 3: Components of the rainwater harvesting system (Source: http://www.clareando.com.br/interno.asp?conteudo=solucoes)

To make up the system, water treatment devices can be inserted, such as water filtration and disinfection. The quality of the water collected will depend on the purpose for which it is used.

According to Mano and Schmitt (2004), rainwater can be used partially or completely. Partial use includes specific applications, such as using the water only for washing vehicles. Total use covers the supply of water for human consumption, personal hygiene, food, among other applications.

The treatment of rainwater will depend on its final destination. Water used for non-potable purposes does not require lengthy purification processes. For simple treatment, natural sedimentation, simple filtration and chlorination processes can be used. For captured water that will be used for human consumption, more complex treatments are recommended, such as ultraviolet disinfection or reverse osmosis (MAY & PRADO, 2004).

Rainwater, used as an alternative source for consumption, is cited by Mierzwa et. al (2007), Seeger et al. (2007), May (2004), Kobiyana et al. (2002), Tomaz (2003), as an action that minimizes the environmental problem caused by water scarcity, even for non-potable uses.

Many authors have commented on the benefits of installing rainwater harvesting systems to reduce flooding. Tucci (2007) considers that rainwater harvesting can be useful as a source of non-potable water supply, but warns that it contributes little to flood control.

Rebouças (2004) comments that regardless of whether water is scarce or abundantly available in certain regions, it should be used sparingly, always assessing the possibility of obtaining it from alternative sources.

2.6 Storage tank

The rainwater storage reservoir has the function of retaining and accumulating the water collected. Sizing depends on three main factors: the demand to be met, the catchment area and the rainfall characteristics of the location (ANA, 2004). According to Tomaz (2005), the sizing of the reservoir for a rainwater harvesting system represents the largest part of the investment and is the main component of the system, since it determines technical and economic viability.

In many cases, the rainfall that occurs at the site will not always be sufficient to meet the required demand, nor will it always be possible to store all the rainfall for economic and spatial reasons. It is therefore essential that the size of the reservoir is compatible with production x demand.

Reservoirs can be supported, elevated or buried, and there are a variety of materials for their manufacture, including reinforced concrete, reinforced block masonry, plastic, polyester, PVC, ordinary brick masonry, among others. However, certain precautions are necessary to maintain the quality of the water to be stored. The ANA/FIESP & SindusCon-SP manual (2005) sets out some of the construction characteristics that reservoirs must respect and some of the precautions to be taken, including:

- Prevent sunlight from entering the tank to reduce the proliferation of algae and microorganisms;
- Keep the inspection cover closed;

- Place a grid or screen at the outlet end of the overflow pipe to prevent small animals from entering;
- Carry out annual cleaning of the reservoir, removing sediment;
- Design the storage tank with a slope at the bottom in the direction of the drainage pipe to facilitate cleaning;
- Ensure that the water collected is only used for non-potable purposes.

There are several methods for sizing reservoirs available in the literature, including those indicated in NBR 15527:2007, and they should be analyzed according to the characteristics of the site (BRASIL, ABNT, 2007).

2.7 Quality of water abstracted for non-potable purposes

The adoption of maintenance and cleaning practices for the rainwater collection system, the insertion of grids and filters and the disposal of the first moments of rain, all contribute to the quality of the water that will be stored. Other equipment can be used to improve the quality of the water, but this will depend on the purpose for which the water is to be used, as it could lead to the system becoming unviable due to the costs involved.

The region where the catchment system is installed can contribute to the quality of the water and its recommended use (MAESTRI, 2003). Philippi et al. (2006) emphasize that several factors influence the quality of rainwater and among these are: the geographical location of the catchment area (proximity to the ocean, urban or rural areas), the presence of vegetation, the presence of pollutant load and the composition of the materials that make up the catchment and storage system (roofs, gutters and reservoir).

The quality of rainwater varies both with the degree of air pollution and with the cleanliness of the catchment system. Tomaz (2003) comments that in highly urbanized areas and industrial centers, pollutants in the air, such as sulphur dioxide (SO_2), nitrogen oxides (NO_X) or even lead, zinc and others, cause changes in the natural concentrations of rainwater.

GROUP RAINDROPS (2002) apud SILVA (2007) presents in Table 1 the applications of water according to where it is collected and its quality.

Table 1. Water quality according to collection site.

Degree of quality	Rain collection site	Observations
A	Roofs (places not occupied by people or	Washing toilets, watering plants, it can be

	animals).	used for human consumption if purified.
B	Roofs (places frequented by people and animals).	Only for non-potable uses (flushing toilets, watering plants), after minor treatment, it cannot be used for drinking.
C	Terraces and waterproofed areas; Parking areas.	Treatment is necessary even for non-potable uses.
D	Roads; elevated railroads.	Even for non-potable uses, it needs treatment.

Source: GROUP RAINDROPS, 2002 apud SILVA, 2007

a. Rainwater Quality Standard

The quality of rainwater for use is a very important issue as each use requires different quality parameters, noting that when there is human contact with the water, the health of the individual must be preserved (KINKER, 2009).

In view of the possibilities for using rainwater, GROUP RAINDROPS (2002) apud SILVA (2007) presents treatment recommendations for the different uses of water in Table 1.

Table 1. Different levels of water quality required according to use

Use of rainwater	Water treatment
Garden irrigation	No treatment is necessary.
Irrigators, firefighting, air conditioners	Care must be taken to keep the equipment in good condition.
Decorative aquatic systems such as ponds, fountains, fountains, mirrors and waterfalls, sanitary flushes in bathrooms, washing clothes and washing cars	Hygienic treatment, due to the possibility of water coming into contact with people.
Bathing/pooling, human consumption and food preparation	Disinfection, as the water is ingested directly or indirectly.

Source: GROUP RAINSDROPS, 2002 apud SILVA, 2007.

If rainwater is used for drinking purposes, it will need to be treated to reach the level of potability set out in the Ministry of Health's Ordinance No. 518/2004, which is valid for Brazil.

2.8 Using rainwater to wash vehicles

The use of rainwater for washing vehicles must have specific characteristics and must not: smell bad; be abrasive; stain surfaces; contain salts or substances left over after drying;

cause infections or contamination by viruses or bacteria harmful to human health (SINDUSCON, 2005).

In many European countries, Japan and the United States, car washing is regulated by specific legislation for small to large businesses. In Brazil, however, this awareness is not yet practiced.

The amount of water used to wash vehicles depends on the equipment used for this, such as machines, high-pressure hoses, hoses, among other alternatives available on the market. Braga (2005) states that approximately 150 liters of water are used to wash a vehicle. According to Morelli (2005), approximately 250 liters of water are used to wash each car and approximately 600 liters of water per urban transport bus.

2.9 Economic Analysis

In addition to environmental issues, the economic potential of using rainwater is an important factor to be verified and analyzed, since this will make it possible to determine whether the system will be financially viable or not.

The economic analysis is based on an estimate of costs resulting from the implementation of the entire system: materials and equipment, labor and electricity costs due to the need to pump the water.

Another issue that must be measured, and which will influence the investment decision, is the determination of the costs related to the consumption of drinking water from the public supply, not considering the use of rainwater. This consumption is analyzed using the water tariff of the concessionaire responsible for the local water supply.

The composition of the water tariff comes from 3 aspects: operating and maintenance costs, economic costs and environmental effects (OECD, 2010). Dornelles (2012) points out that this composition is the sum of operating and maintenance costs, costs caused by the increase in conflicts over water use and the degradation of its quality, the possible effects of climate change and the poor management of water resources.

In view of this, it is expected that the trend in water tariffs will gradually increase as a result of the factors mentioned above and, at the same time, it is expected to stimulate the implementation of rainwater collection and use systems by reducing the return on investment.

a. Water and sewage tariffs in the municipality of Pelotas

Municipal Law No. 6.294 of 2015 establishes criteria for charging SANEP (Autonomous

Sanitation Service of Pelotas) for the supply of water, collection and treatment of effluents, and the tariff structure is divided into categories and according to consumption measured per cubic meter.

The categories are divided into I - Residential; II - Social Residential; III - Philanthropic; IV - Commercial or Service; V - Industrial; VI - Public and VII - Religious Temples and Houses of Religion. The enterprise under study is classified in category IV as a service provider.

According to article 6 of this municipal law, the sewage generated is charged according to the distribution of the effluent collection network in the municipality of Pelotas:

- Effluent collected and treated - 80% of the water tariff charged;
- Effluent collected and removed (untreated) - 60% of the water tariff charged;
- Effluent discharged directly into the drainage system - 30% of the water tariff charged;
- Effluent discharged directly into the drainage system when SANEP does not provide maintenance and clearance services - No charge.

Table 3 shows the tariff charged for water supply and sewage collection.

Table 2. Tariff matrix for the municipality of Pelotas - RS.

Category IV: Commercial/Services				
Basic Service	R$ 32,35			
Base Price (R$/m³)	**Water**		**Sewage**	
	-	30%	60%	80%
Up to 10 m³	R$ 4,35	R$ 1,31	R$ 2,61	R$ 3,48
From 11 m3 to 15 m3	R$ 5,00	R$ 1,50	R$ 3,00	R$ 4,00
From 16 m3 to 20 m3	R$ 5,75	R$ 1,73	R$ 3,45	R$ 4,60
From 21 m3 to 30 m3	R$ 6,62	R$ 1,99	R$ 3,97	R$ 5,30
From 31 m3 to 50 m3	R$ 7,61	R$ 2,28	R$ 4,57	R$ 6,09
From 51 m3 to 100 m3	R$ 8,75	R$ 2,63	R$ 5,25	R$ 7,00
Above 100 m3	R$ 10,07	R$ 3,02	R$ 6,04	R$ 8,06

Source: Municipal Law No. 6.294/2015.

The installation of rainwater harvesting systems reduces and saves water consumption from the supply network, but does not reduce the generation of sanitary sewage. Dornelles (2012) comments that regardless of the source of water (piped, well, water truck or rainwater), the sewage generated should be charged for.

In view of this, Municipal Law No. 6,294 of 2015 establishes in its Article 8:

> Art. 8 Properties that are connected to the public sewage system and have their own sources of supply, or make use of rainwater, must also have water metering from the alternative source, with the aim of billing and collecting the volume of sewage produced.

Thus, in addition to the conventional metering of the water supply by the concessionaire, it is necessary to install a measuring register for the volume of water that will be captured by the rainwater harvesting system, in order to measure the volume of water that will be discharged into the sewage network for the properties that are connected to it.

3. METHODOLOGY

3.1 Procedures

Initially, in order to draw up the rainwater harvesting project, studies were carried out and alternatives surveyed for its development, through books, standards, legislation, manuals and technical papers related to the central theme of the work, in order to use the water in the vehicle washing and cleaning sector of a passenger road transport company and also to clean the floors in the maintenance sector of the enterprise.

The necessary information and data about the company, such as the number of vehicles washed per day, the volume of water needed for each wash, the equipment used in the operation, the days of the week on which the sector operates, the area of the company's roof and measurements of the existing structures, were collected by visiting the site and also by talking to the person in charge of the company's maintenance sector.

Once the information was available, the sizing of the reservoir and pumping system was carried out, along with a survey of costs related to consumption and predicted savings in drinking water, in order to verify the economic viability of the rainwater collection system for the project.

3.2 Characterization and study area

The municipality of Pelotas is located in the extreme south of the state of Rio Grande do Sul - RS, on the banks of the Sâo Gonçalo Channel, which connects the Patos Lagoon and the Mirim Lagoon, the latter being the largest in Brazil. It has latitude 31°46'34" S, longitude 52°21'34" W (Coordinate System: WGS - 84 Fuso 22S; Projection: Universal Mercator Traverse) and an average altitude of 9 m. It has a total area of approximately 1610 km^2, as shown in Figure 4.

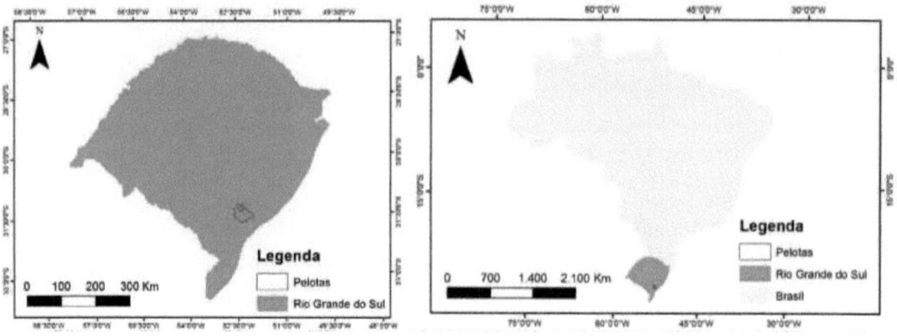

Figure 4 - Location of the municipality of Pelotas - RS.

The work was carried out at Expresso Embaixador, an intercity passenger transportation company, located at Rua Giuseppe Garibaldi n° 660, Centro, in the municipality of Pelotas - RS, specifically at coordinates 31°77'66" S and 52°33'03" W (Coordinate System: WGS - 84 Fuso 22S; Projection: Universal Traverse de Mercator), as shown in Figure 5. The project covers 13,000 m^2, where the administrative and operational activities take place.

Figure 5 - Map of the location of the Expresso Embaixador company in the municipality of Pelotas - RS.

The company currently has a fleet of 100 vehicles, which carry out intercity passenger transportation in various municipalities in the southern region of the state. The vehicle washing sector operates with two washing machines and high-pressure hoses, and 95 vehicles are washed every day of the week. Figure 6 shows how the external washing process works.

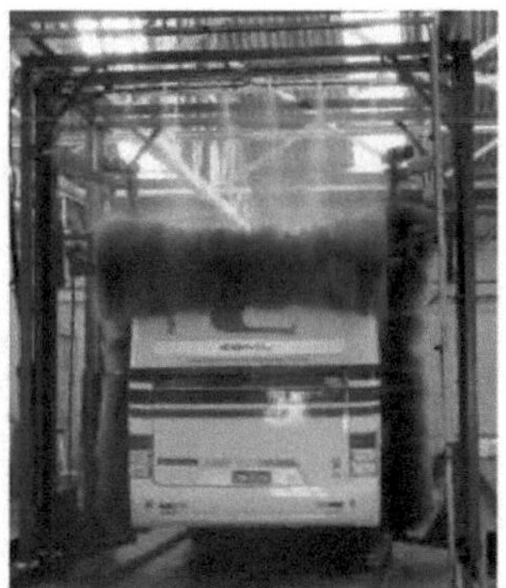

Figure 6. External washing process at Expresso Embaixador.

In addition to the external washing, the vehicles are also cleaned internally and the headboards (seat covers) are washed. The internal cleaning is carried out manually by employees and the headlining is washed by a washing machine.

3.3 Sizing the rainwater collection system

3.3.1 Consumption forecast

The demand for water needed to clean the vehicles was checked according to the volume of water used to operate the washing machine, high-pressure hose, internal cleaning, washing the headboards (seat covers) and cleaning the floor in the company's maintenance department.

Washing begins with the use of a high-pressure hose for an average of 5 minutes, which has a flow rate of 26 l.min^ 1. The vehicle then goes into the washing machine, which needs 300 liters of water to clean each vehicle. 95 buses are washed every day. For internal cleaning, an average of 3 liters of water is used for each vehicle. The headland washing machine consumes 440 liters of water per wash. Six washes are carried out every day, totaling 44,200 liters of water per day.

The maintenance floor is cleaned weekly using high-pressure hoses (26 $l.min^{-1}$) for approximately 10 minutes.

Table 4 shows the daily, monthly and annual demands for carrying out the activities described.

Table 3. Demand required to wash vehicles internally and externally, headboards and clean the floor in the maintenance department.

Demand	Unit	Water volume
Daily	$m^3.day^{-1}$	44,46
Monthly	$m^3.month^{-1}$	1.333,86
Annual	m3.year-1	16.006,37

3.3.2 Contribution area

Rainwater will be collected from the company's roofs, which include the garage (6,278 m^2), warehouse (900 $m^{2)}$, washing (1,402.5 m2) and mechanical workshop (1,964 m2). As this is a large area of roofs, it was decided to run simulations with several areas in order to check the most viable alternative. Appendices A, B and C show a sketch of the plan of the development and the location of the respective catchment areas.

Table 5 shows the options for carrying out the simulations.

Table 4: Simulations of rainwater collection areas and their respective materials.

Simulation	Sectors	Roof type	Total area (m^2)
1	Garages	Metal and ceramic tiles	6.278
2	Garages + washing + storage	Metal and ceramic tiles	8.580,5
3	Garages + Laundry + Storage+ maintenance	Metal and ceramic tiles	10.544

In this context, according to Tomaz (2003), the *runoff* coefficient, indicated by the letter "C", varies according to the composition of the roof material from 0.8 to 0.90, and the closer it is to 1, the greater the water runoff. This water loss is taken into account as a result of cleaning the roof, evaporation loss and so on. Table 6 shows the variation in the coefficient according to the roof material.

Table 5. Average runoff coefficients

Material	Runoff coefficient
Ceramic tiles	0,8 a 0,9
Glazed tiles	0,9 a 0,95
Corrugated metal roof tiles	0,8 a 0,9
Asbestos cement	0,8 a 0,9
Plastic, PVC	0,9 a 0,95

Source: Tomaz (2003)

3.3.3 Rainfall characteristics

Precipitation data for the municipality of Pelotas was obtained from the Pelotas Agroclimatological Station (EAP), located in the municipality of Capâo do Leâo (RS), 7.5 km from the center of Pelotas - RS. The EAP is operated under an agreement between Embrapa Temperate Climate, the Federal University of Pelotas and the National Institute of Meteorology.

The rain gauge is located at latitude 31°48'12" and longitude 52°24'41" and altitude 13.24 m (EAP, 2016). This point is 8.25 km away from the Expresso Embaixador company. The data covers 36 years of monitoring (1980 to 2015) and was chosen because it represents the most recent flooding and drought events in the municipality.

The rainfall data series is daily, so for the purposes of water runoff over the catchment surface, data of less than 1 mm was not taken into account, and when less than this parameter, it was considered as no rainfall at the site.

Appendices E and F show the average values of monthly and annual rainfall, the months in which rainfall was less than 100 mm and the longest period of consecutive days without rainfall in the municipality of Pelotas during the study period (1980 to 2015).

3.3.4 Storage tank

The water storage tank must be made of corrosion-resistant material or have an anti-corrosive coating on the inside (NBR 5626/98).

Currently, there are various methods for sizing rainwater tanks in the literature, in technical papers and in the NBR 15527 standard (ABNT, 2007). Some of these methods were applied and analyzed in this work in order to select the feasible reservoir volume according to the spatial, technical and economic conditions of the project.

a. Rippl method

Tomaz (2005) points out that this method presents the extreme value of the reservoir volume and it is important to obtain it as a maximum reference, compared to the other volumes obtained by applying the other methods.

The methodology is based on the historical series of daily or monthly rainfall, the catchment area and water consumption. The sum of the volume of rainfall and demand over the same period will determine the excess or lack of water available in the reservoir. The accumulated volumes determine the maximum volume reached, which must be taken into account when sizing the reservoir, as can be seen from equations 1, 2 and 3.

$$Q_{(t)} = \frac{P_{(t)} * A * C}{1000}$$

Equation 1

Being:

$Q_{(t)}$ Volume of rainfall captured at time t (m^3);

$P_{(t)}$: Precipitation at time t (mm);

A: Catchment area (m^2);

C: Runoff coefficient (dimensionless).

$$S_{(t)} = D_{(t)} - Q_{(t)}$$

Equation 2

Being:

$S_{(t)}$: Volume of water in the reservoir at time t (m3);

$Q_{(t)}$ Volume of rainfall captured at time t (m3);

$D_{(t)}$: Demand or consumption at time t (m3).

$$V = \sum S_{(t)},$$ only for values $S_{(t)} > 0$ Equation 3

The roofs of the development are made of different materials, which vary according to the buildings analyzed. In view of this, and analyzing Tables 4 and 5, the runoff coefficient selected was C= 0.85.

b. Simulation Method

This method consists of evaluating the water balance within the reservoir, defining a volume for the reservoir and verifying the need for an external water supply to meet demand or the occurrence of water overflow from the reservoir. According to Tomaz (2005), in order to carry out these calculations, it is necessary to assume that the historical data available is representative of future conditions.

According to NBR 15527 (2007), water evaporation is not taken into account when applying this method.

Initially, the volume of precipitation that reaches the catchment surface is calculated using equation 4.

$$Q_{(t)} = \frac{P_{(t)} * A * C}{1000}$$

Equation 4

Being:

$Q_{(t)}$ Volume of rainfall captured at time t (m^{3});

$P_{(t)}$: Precipitation at time t (mm);

A: Catchment area (m^{2});

C: Runoff coefficient (dimensionless).

Then, using equation 5, the water balance inside the reservoir is calculated in order to verify the occurrence of overflow or the need for an external water supply. It starts with the reservoir at zero.

$S_{(t)} = S_{(t-1)} + Q_{(t)} - D_{(t)}$ Equation 5

Being:

$S_{(t)}$: Volume of water in the reservoir at time t (m3);

$S_{(t-1)}$: Volume of water in the reservoir at time t-1;

$Q_{(t)}$ Volume of rainfall captured at time t (m3);

$D_{(t)}$: Demand or consumption at time t (m3).

It is important to analyze the data resulting from each operation, and when these are negative, it means that there was a need to supply water from an external source, i.e. the volume of water abstracted was less than the project's water consumption. To define the volume of the reservoir, it is necessary to find a balance between the water overflowing from the system and the need for external supply. The efficiency of the system can be calculated using equation 6.

$$E_f = \frac{N}{Nr}$$

Equation 6

Being:

E_f : System efficiency

N: Number of months in which the reservoir is able to meet demand;

Nr: Total number of months in the period analyzed.

Efficiency can result in a very low value, in which case another volume value is adopted for the reservoir so that the method is interactive, until the interests of the project (cost-benefit ratio) are satisfied.

c. Brazilian Practical Method

This method is also known as the Azevedo Neto Method. Its methodology is practical and consists of manipulating the average annual rainfall and also the sum of months with "little rain", calculated by equation 7.

For the study, the criteria adopted to characterize the "little rain" parameter were dry months (no rain) and months in which rainfall was less than 100 mm.

$V = 0.0042 * P * A * T$ Equation 7

Being:

V: Reservoir volume (m^3);

P: Average annual rainfall (mm);

T: Number of months of "little rain" or drought during the year;

A: Catchment area (m^2).

d. English Practical Method

This method is simple and is based on an ideal reservoir volume of 5% of the average annual rainfall collected, with demand being disregarded. The reservoir volume to be used is the largest resulting value and can be calculated according to equation 8.

$$V = 0.05 * A * P \qquad \text{Equation 8}$$

Being:

V: Reservoir volume (m^3);

P: Average annual rainfall (mm);

A: Catchment area (m^2).

e. German Practical Method

As described by NBR 11527 (2007), it is an empirical method, where the parameter with the lowest value is applied: 6% of the annual volume of consumption or 6% of the annual volume of rainfall captured (Equation 9).

$$V_{adopted} = \min (V_c; D) * 0.06 \qquad \text{Equation 9}$$

Being:

$V_{adopted}$: Reservoir volume (m3);

V_c: Annual volume of rainwater collected (m3);

D: Annual non-potable water demand (m3).

f. Australian Practical Method

This method uses average monthly rainfall data, the catchment area and the necessary demand, as described in equation 10.

$$Q=A*C*(P-I) \qquad \text{Equation 10}$$

Being:

Q: Monthly volume of rainfall captured (m3);

A: Catchment area (m2);

C: Runoff coefficient (dimensionless);

P: Average monthly rainfall (mm);

I: Losses (first rain, evaporation, etc.) (mm).

According to NBR 15527 (ABNT, 2007), the calculation of the volume of the reservoir is carried out by trial and error until optimized values of reliability and volume are used, using equations 11, 12 and 13.

$$V_{(t)} = V_{(t-1)} + Q_{(t)} - D_{(t)}$$

Equation 11

Being:

$V_{(t)}$ Volume of water in the reservoir at the end of month t (m3);

$V_{(t-1)}$ Volume of water in the tank at the start of the month ($m^{3)}$;

$Q_{(t)}$: Monthly volume produced by rainfall in month t;

$D_{(t)}$: Monthly demand (m3).

For calculation purposes, assume that the tank is empty in the first month.

When $(V_{((t-)(1))} + Q_{(()(t)())} - D_{(()(t)())}) < O$, then $V_{(()(t)())} = 0$ Equation 12

Once the volumes have been obtained, the reliability is calculated in order to verify the efficiency of the reservoir, as shown in equation 13.

$$P_r = \frac{N_r}{N}$$

Equation 13

Being:

P_r: Failure;

N_r : Number of months in which the reservoir did not meet demand (V(t) = 0);

N: Number of months considered.

Confidence = 1 - P_r Equation 14

The method recommends that confidence values be between 90% and 99%.

g. Method of consecutive days without rain

Group Raindrops (2002) presented this method using historical rainfall data which shows the number of consecutive days without rain and the water demand required (equation 15).

$V = C_d * D_{sc}$ Equation 15

Being:

V: Reservoir volume (m3);

C_d: Daily water consumption (m3);

D_{sc}: Consecutive days without rain.

However, in order to know the consecutive days without rain, it is necessary to make a statistical adjustment to the historical rainfall series analyzed, as described below:

- It first counts the number of consecutive days without rain for each year of the historical series, in this case rainfall of less than 1 mm.
- Using the Excel program, these data were arranged in descending order, keeping the highest order for repeated data and calculating the probability using equation 16.

$$P = \frac{m}{(n+1)}$$

Equation 16

Being:

Q: Probability;

m: order number;

n: total number of historical series analyzed.

- The payback period (T_r) is calculated using the relationship expressed in equation 17.

$$Tr = \frac{1}{P}$$

Equation 17

- Gumbel's Law transforms the return period into a reduced variable by applying equation 18.

$$b = -ln\left[ln\left(\frac{Tr}{Tr-1}\right)\right]$$

Equation 18

Once the data is available, a linear adjustment is made and the return period (Tr) for 10 years is applied in order to check the drought period to be applied in equation 15.

3.4 Water pumping

The water collected will be conveyed by gravity to the first reservoir, from where, by installing a hydraulic pump, the water will be conveyed to another reservoir for later use.

Standard NBR 12214 (ABNT, 1992), entitled "Design of a water pumping system for public supply", sets out the conditions for drawing up these designs.

To carry out these projects, it is necessary to observe the recommendations for the suction and discharge pipes, heights and flow rates for the motor-pump set.

After selecting the power of the motor pump and checking the respective flow rate (m^3h^{-1}) for its operation, the daily operating time and the number of days of use in the month were estimated. To determine the costs of electricity due to pumping, we used the data on the motor pumps adopted and the values (R$.$kWh^{-1}$) charged by CEEE for the category in which the project falls.

According to Marinoski (2007), equation 19 was used to determine the consumption of electricity spent on the pumping process. In this study, taxes charged by the energy supply company were not added.

$$C_{EE} = P_{MB} * t * V_{CEEE}$$

Equation 19

Being:

C_{EE}: Monthly cost of electricity for pumping rainwater (R$);

P_{MB} : Motor pump power (kW);

t: Motor pump operating time ($h.day^{-1}$);

$V_{(CEEE)}$: Amount charged by CEEE for electricity consumed (R$.$kWh^{-1}$).

a) CEEE charges for electricity

The amount charged by CEEE (Companhia Estadual de Energia Elétrica) was verified according to the tariff currently charged to the company, however this tariff is not fixed due to the rates assigned according to monthly consumption.

In this way, an average of the peak and off-peak active consumption tariffs was calculated, which correspond to the times when the energy is used. The average tariff resulted in a value of R$1.05 kWh, without the addition of flags (green, yellow and red).

Therefore, for calculation purposes, the fixed value of R$ 1.05 kWh will be used.

3.5 Economic viability

a) Costs of water and sewage discharged in the municipality (before the installation of the rainwater harvesting system)

In order to estimate the amount that would be spent on the public water supply to carry out the vehicle washing and cleaning services, if the development did not install a rainwater collection system, it was calculated using 3 equations, taking into account the parameters described in municipal legislation No. 6.294 of 2015, which describes the tariff matrix charged for the supply and collection of sewage generated.

Initially, the average monthly volume of water consumed to cover the company's activities (internal and external washing of vehicles, laundry and cleaning of the maintenance sector) was calculated, according to equation 20.

$$V_{AC} = (V_{máq} + V_{mag} + V_{lav} + V_{LI} + V_{man})$$

Equation 20

Being:

V_{AC}: Average monthly volume of water consumed ($m^3.month^{-1}$);

$V_{machine}$: Average monthly volume of water used by the washing machine ($m^3.month^{-1}$);

V_{mag}: Average monthly volume of water used by the hose ($m^3.month^{-1}$);

V_{lav}: Average monthly volume of water used in the laundry ($m^3.month^{-1}$);

V_{LI}: Average monthly volume of water used for internal vehicle washing ($m^3.month^{-1}$);

V_{man}: Average monthly volume of water used to clean the maintenance sector ($m^3.month^{-1}$).

Using the average monthly volume of water consumed, we calculated the average monthly cost for water supply and sewage disposal, expressed in equation 21.

$$C_{Ab} = T_{SB} + (T_{VAC} * V_{AC}) + (V_{AC} * P_{EG})$$

Equation 21

Being:

C_{Ab}: Average monthly cost of water supply (R$.$month^{-1}$);

T_{SB}: Basic Service Charge (R$ $month^{-1}$);

T_{VAC}: Rate of the average volume of water consumed (R$.$m^{-3}$);

V_{AC}: Average monthly volume of water consumed (m3.$month^{-1}$);

P_{EG}: Percentage of sewage generated according to the distribution of the municipality's sewage network (R$.m^{-3}).

b) Costs of water and sewage discharged in the municipality (after installation of the rainwater collection system)

After the rainwater harvesting system has been installed, there will be days when the water stored in the reservoirs will not be enough to meet the company's demand. In this sense, and in order to continue carrying out water-dependent activities, it will be necessary for the municipal water supply to complement the system.

To quantify the average water supply, it will be necessary to analyze the water balance inside the reservoir, as proposed by the simulation method.

$$C_B = T_{SB} + (T_{VAC} * V_{AS}) + (V_{AS} * P_{EG})$$

Equation 22

Being:

C_B: Average monthly cost of drinking water to supply demand (R$.month^{-1});

T_{SB}: Basic Service Charge (R$ month^{-1});

T_{VAC}: Rate of the average volume of water consumed (R$.m^{-3});

V_{AS}: Average monthly volume of water needed to supply demand (m^3.month^{-1});

P_{EG}: Percentage of sewage generated according to the distribution of the municipality's sewage network (R$.m^{-3}).

Equation 22 shows the new average monthly cost of drinking water that the company would pay after installing the rainwater harvesting system.

c) Savings after installing rainwater harvesting system

To check economic viability, and in accordance with the methodology proposed by Marinoski (2007), the difference was made between the average monthly cost of water from the public supply (before the system was installed) and the average monthly cost of water (after the system was installed) plus the cost of the electricity needed to run the system, as shown in equation 23.

$$E = C_{Ab} - (C_B + C_{EE})$$

Equation 23

Being:

E: Average monthly savings after installing the rainwater harvesting system (R$.month^{-1});

C_{Ab}: Average monthly cost of water supply (before system installation) (R$.month^{-1})

C_B: Average monthly cost of drinking water to supply demand (after system installation) (R$.month^{-1});

C_{EE}: Average monthly cost of electricity for pumping water into the system (R$.month^{-1}).

4. RESULTS AND DISCUSSION

4.1 Potential use of rainwater in Pelotas

The 36-year period analyzed for precipitation in the municipality of Pelotas, from 1980 to 2015, includes a total of 13,148 days of verification, of which approximately 72.8%, or 9,571 days, did not record any precipitation.

Figure 7 shows the total annual rainfall in the municipality of Pelotas over the period analyzed, highlighting the annual municipal and regional average.

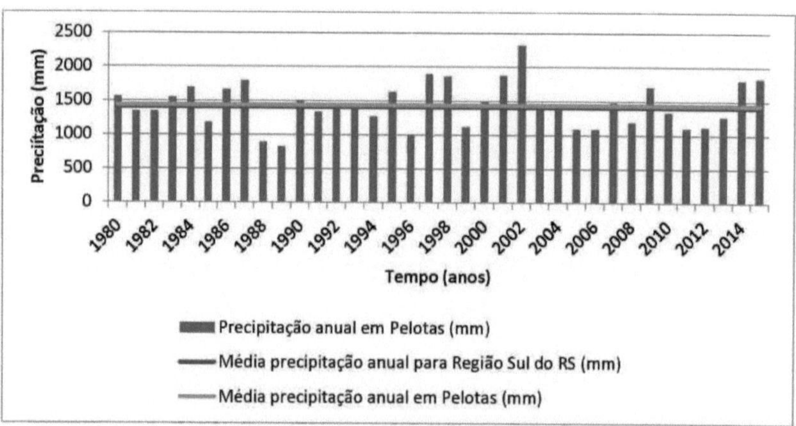

Figura 7. Annual precipitation in the municipality of Pelotas from 1980 to 2015.

The average rainfall for the southern region of the state of Rio Grande do Sul is between 1299 mm and 1500 mm. For the period analyzed, the municipality of Pelotas had an annual average of 1438 mm, with a maximum of 2315 mm in 2002 and a minimum of 823 mm in 1989, as can be seen in Figure 8. In view of this, it can be seen that the distribution of rainfall is practically regular over the period analyzed, with a standard deviation of approximately 320 mm between the maximum and minimum rainfall recorded.

Figure 8 shows the monthly averages for the municipality of Pelotas during the period analyzed, together with the state average for the southern region of the state.

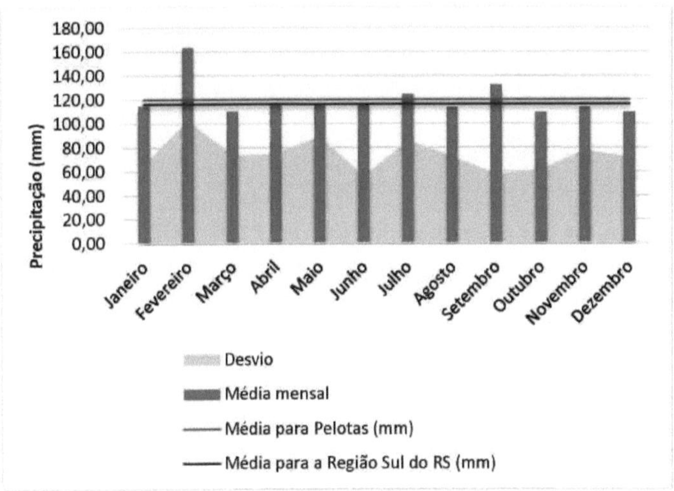

Figura 8. Average monthly rainfall in the municipality of Pelotas from 1980 to 2015.

Looking at Figure 8, it is possible to see that the monthly distribution of rainfall in the municipality of Pelotas is regular, with no major variations in the different seasons of the year, except for February, where it was higher than in the other months. However, when the monthly standard deviation of the data is analyzed, it can be considered that these rainfall results were isolated cases, leading to an increase in the average.

This analysis of rainfall in the municipality provides greater security for the sizing of the storage reservoir for the captured rain, since, as commented by Cohim et al. (2008), high rainfall indices and more constant rainfall distributions throughout the year allow for the use of smaller reservoir volumes.

Figures 9, 10 and 11 show the rainwater collection capacity, considering the 3 area size simulations (6,278 m^2, 8,580.5 m^2 and 10,544 $m^{(2)}$), the runoff coefficient according to the company's roof material and the rainfall in the municipality of Pelotas over the 36-year period, together with the average daily collection for the 3 cases.

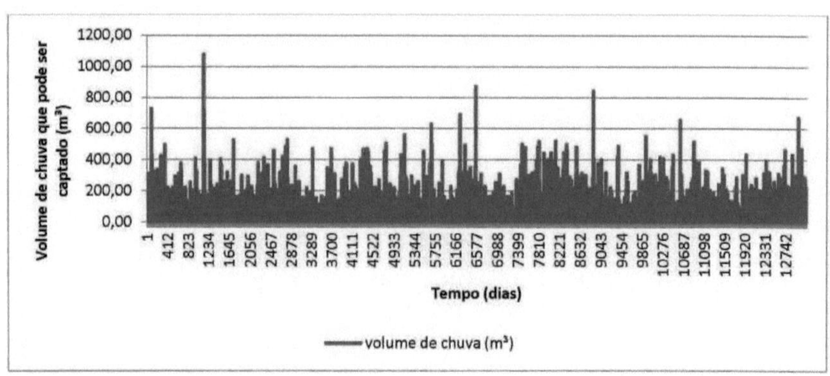

Figure 9. Volume of rain possible to capture for simulation 1 (6,728 m^{2}).

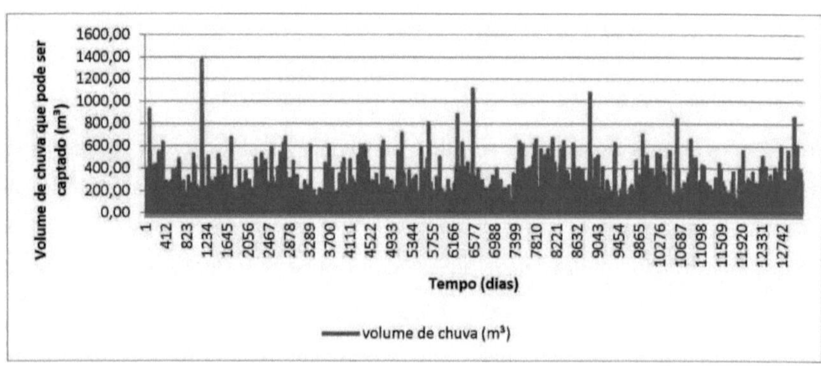

Figura 10. Volume of rain possible to capture for simulation 2 (8,580.5 m2).

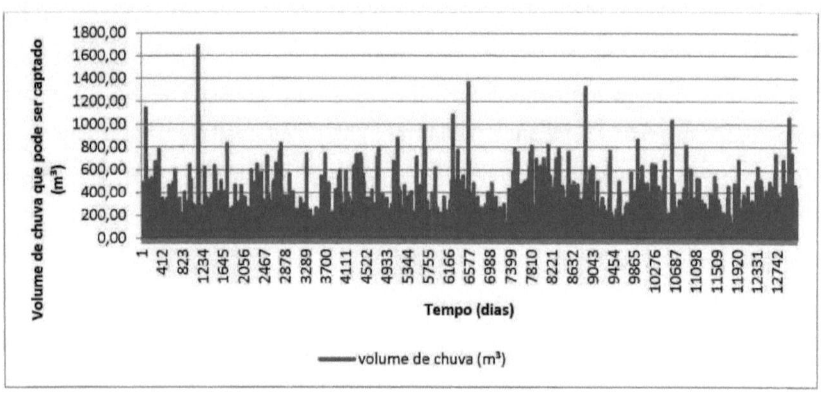

Figura 11. Volume of rain possible to capture for simulation 3 (10,544 m2).

As can be seen, and also commented on by Carvalho et al. (2007), the size of the catchment area influences the volume of water captured, i.e. the larger the area, the greater the volume of usable rainfall.

It is also possible to observe that the peaks in the volume of rainwater that can be captured in the 3 situations are generally isolated cases, indicating that the adoption of large reservoirs to store these volumes will not always be the most advisable and applicable to the project, due to the availability of space for the installation of the reservoir.

The increase in the average daily catchment volume of simulation 1 (Figure 9) compared to simulation 2 (Figure 10) is approximately 37%. The increase in simulation 2 (Figure 10) compared to simulation 3 (Figure 11) is approximately 23%. This increase in the average water withdrawal refers to the decrease in reservoir volume in each method applied, since the water demand required to meet the company's activities remains constant for the 3 simulations carried out.

However, in agreement with Cohim (2008), the greater the annual rainfall, the greater the volume of water that can be captured, but its use depends on the size of the reservoir that will be adopted.

4.2 Sizing the rainwater storage tank

By manipulating the rainfall data (monthly, annual and consecutive days without rain), the variation in the catchment coverage area and the water demand required to carry out the company's activities, applied to the methods (Rippl, Consecutive days without rain and Brazilian, German and English practices) for sizing the water storage reservoir, multiple volumes were obtained, which can be seen in Table 7.

Table 6. Results of reservoir volumes using sizing methods.

Method	Maximum reservoir volume (m^3)		
	Simulation 1	**Simulation 2**	**Simulation 3**
Brazilian Practice	290	370	455
German Practice	66	84	104
Practical English	65	83	102
Rippl	7785	5522	2946
Consecutive days without rain	1457	1457	1457

Figure 12 shows a summary of the results of the reservoir volumes for each simulation carried out and method applied. It can be seen that the more practical the methodology used in each sizing method, the smaller the variation in volume for the reservoir.

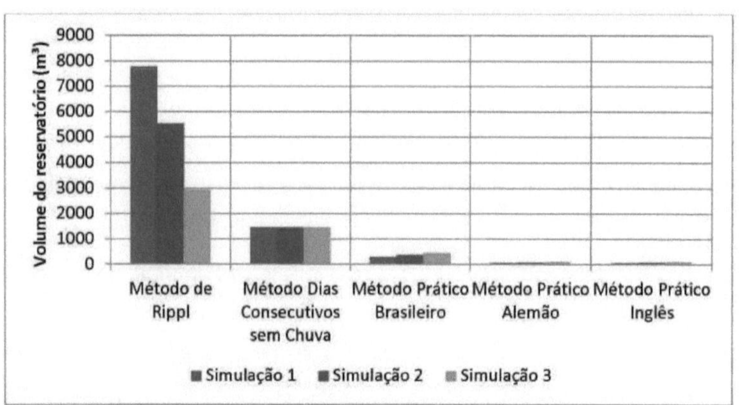

Figure 12. Reservoir volume for each simulation applied.

The reservoir volumes varied greatly from one to another. Moruzzi and Oliveira (2010) commented in their study on reservoir sizing that the practical methods are generally not very sensitive to variations in water demand and catchment area. This can also be seen from the results presented in Table 7 for the Brazilian, German and English practice methods.

It is important to note that the Brazilian (or Azevedo Neto), English and German practice methods consider average annual rainfall in their calculations, which can undersize or oversize the reservoir, not taking into account El Nino or La Nina events that can occur in the region.

The sizing of the reservoir carried out by applying the German and English practical methods was similar, as they use the same parameters in their formulation, i.e. average annual rainfall and catchment area. The only difference between the methods is that the German method implies using the lowest value between the usable volume and the required demand, and as the latter was always higher, the usable volume data was used, generating proximity between the sizing results.

The demand for water needed to carry out the company's activities is an important parameter to check when sizing the reservoir, since you want it to meet the company's needs most of the time. In this respect, it can be seen that practical methods do not make use of this parameter, which can also compromise the under- or over-sizing of the reservoir.

For the Rippl method, under the same rainfall and water demand conditions, the catchment area is inversely proportional to the final volume of the reservoir applied in each simulation, i.e. the larger the catchment area, the smaller the volume of the storage reservoir. This observation can also be seen in the study carried out by Amorim and Pereira (2008).

To size the reservoir using the Consecutive Days without Rain (DSC) method, it takes into account the longest period of drought in the year, the probability of its occurrence, the 10-year return period and finally the demand for water, resulting in the volume of water reserves required. This method does not take into account catchment coverage, which is why all the results for reservoir capacity remained constant for the 3 simulations.

It can be seen that the Brazilian practice method and the Consecutive Days without Rain method use data from the municipality's drought periods, but use different criteria to verify the intensity of insufficient rainfall. Dornelles et al. (2010), in their study evaluating the sizing methods, for both methods mentioned and under the same average annual rainfall and return period, found higher reserve volumes for the Brazilian method and lower volumes for the DSC, which does not agree with the results found for this study, due to the size of the catchment area used.

Amorim and Pereira (2008) comment that the practical methods, because they are less complex, are recommended for sizing reservoirs for smaller establishments, such as homes, while the more complex methods, such as Rippl, Analysis and Simulation and Consecutive Days without Rain, are recommended for larger projects, which require a greater volume of water in their applications.

As suggested by Carvalho et al. (2007), the Analysis and Simulation Method can be used in combination with other sizing methods, by analyzing the behavior of the reservoir volumes resulting from the methods applied, allowing greater precision in adopting the volume for the reservoir depending on its desired efficiency.

In this context, the volumes calculated for the reservoirs resulting from the application of the Rippl, Brazilian, German, English and consecutive days without rain methods were fixed and applied as volumes for the Australian Practical Method (APM) and Analysis and Simulation Method (ASM), in order to verify the water balance within the reservoir and its efficiency.

For simulation 1, the results applied to the Analysis and Simulation Method and the Australian Practical Method can be seen in Table 8 and summarized in Figure 13.

Table 7. Reservoir sizing for simulation 1 with monthly rainfall data.

	Source method	Reservoir volume	Average monthly supply	Monthly Overflow Average	Efficiency
		m^3	m3. month-1	m3. month-1	%
Australian Practical Method	Rippl	7785	654	0	9%
	DSC	1457	655	0	9%
	MAS/MPA	1000	658	3	9%
	MAS/MPA	500	665	10	8%
	Brazilian	290	670	15	7%
	MAS/MPA	250	671	16	7%
	German	66	678	24	7%
	English	65	678	24	7%
Method Analysis and Simulation	Rippl	7785	649	0	10%
	DSC	1457	650	0	10%
	MAS/MPA	1000	652	3	9%
	MAS/MPA	500	659	10	8%
	Brazilian	290	664	15	8%
	MAS/MPA	250	665	17	7%
	German	66	673	24	7%
	English	65	673	24	7%

From the results shown in Table 8, it can be seen that the difference between the MPA and MAS methods is small, approximately 1%. This difference refers to the calculation methodology used by both methods, with the MPA taking into account a loss due to evaporation and interception of the water before it reaches the reservoir, while the MAS does not.

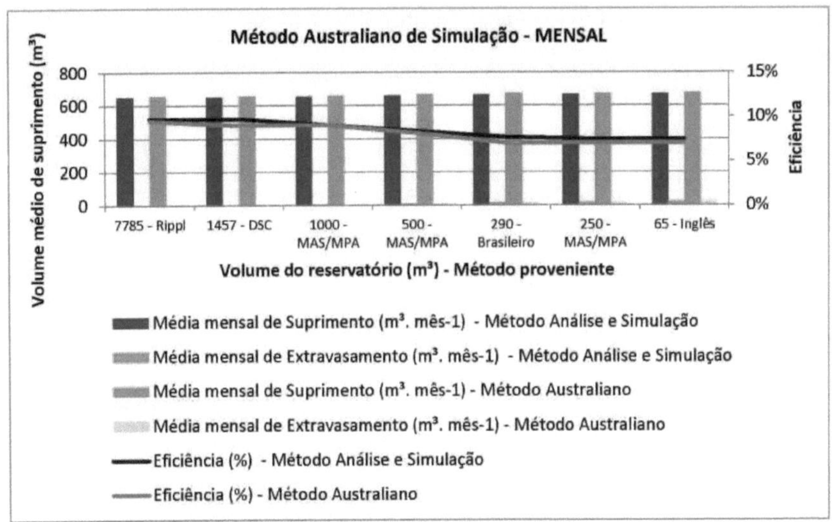

Figura 13. Reservoir sizing for simulation 1 with monthly rainfall data

Checking the water balance in each dimensioned volume, it is possible to see that even applying a reservoir with a capacity of 7,785 m^3 would still require a supplement of approximately 649 m3 of water per month. On the other hand, applying a reservoir with a capacity of 1,000 m3, about 8 times smaller than the first one, the need for external supply would be approximately 652 m3, i.e. only 3m3 more per month. Finally, if a 65 m3 reservoir were used, 120 times smaller than the first and 16 times smaller than the second reservoir, the need for external supply would be approximately 673 m3, 24 m3 more than the first reservoir analyzed and 21 m3 more than the second.

It can also be seen that even using the smallest reservoir dimensioned (65 m3) for this area simulation, the volume of water that overflows from the system is approximately 24 m3 per month. This is because the reservoir does not have the volumetric capacity to store all the precipitation that runs off the roof.

The efficiency for this first simulation, which shows a catchment area of 6,728 m^2 is low, justified by the high demand for water required to carry out the company's activities and also by the natural precipitation factor that occurs in the municipality, which cannot be altered.

In this way, the water balance inside the reservoir for the first simulation would only be more efficient if the municipality's rainfall were higher, the demand for water were lower or if the catchment area were increased.

For simulation 2, which includes 8,580.5 m2 of catchment area, around 22% larger than the first simulation, the results of the volumes for the reservoirs are shown in Table 9 and summarized in Figure 14, applied to the Analysis and Simulation Method and the Australian Practical Method, using monthly rainfall data.

Table 8. Reservoir sizing for simulation 2 with monthly rainfall data.

	Source method	Reservoir volume m^3	Average monthly supply m3. month-1	Average monthly overflow m3. month-1	Efficiency %
Australian Practical Method	Rippl	5522	468	0	25%
	DSC	1457	477	9	23%
	MAS/MPA	1000	486	18	22%
	MAS/MPA Brazilian MAS/MPA	500	502	34	19%
	German	370	508	41	18%
		250	517	49	18%
		84	534	66	17%
	English	83	534	66	17%
Method Analysis and Simulation	Rippl	5522	463	0	25%
	DSC	1457	472	9	23%
	MAS/MPA	1000	480	18	22%
	MAS/MPA Brazilian	500	495	34	19%
	MAS/MPA	370	502	42	19%
	German	250	510	50	18%
		84	527	67	17%
	English	83	528	68	17%

Comparing the results of both methods, it can be seen that the difference between the external water supply results is approximately 1% for the largest reservoir and 3% for the smallest, so the comparison between the reservoir volumes will be made using the smallest result, i.e. MAS.

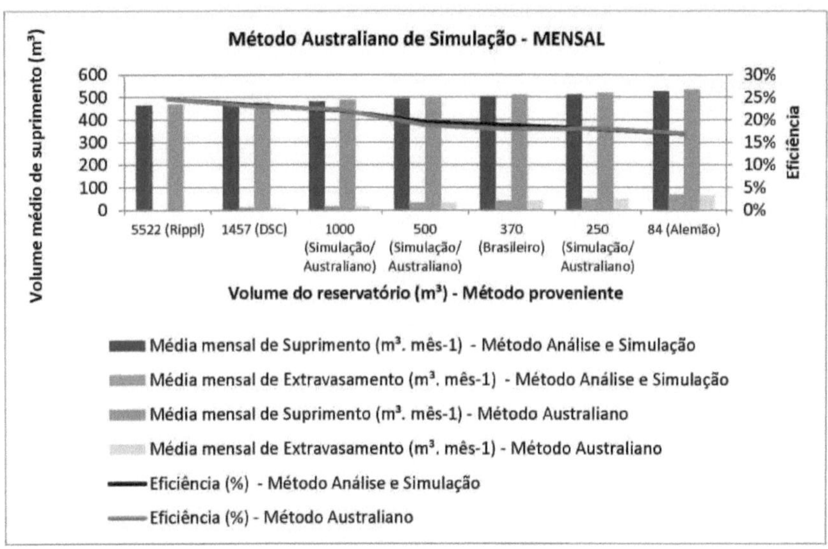

Figura 14. Reservoir sizing for simulation 2 with monthly rainfall data

Checking the resulting volumes for the reservoirs, and choosing the volumes of 1,457 m^3, 500 $m^{(3)}$and 250 m^3 at random, it can be seen that by using a 500 m3 reservoir instead of a 1,457 m3 reservoir, not only is the size approximately 3 times smaller, but the amount of water needed for external supply is around 495 $m3.month^{-1}$, around 5% more water. On the other hand, by using a 250 m3 reservoir instead of a 500 m3 reservoir, not only is the size reduced by half, but the supply requirement increases by only 8% compared to the first reservoir and 3% compared to the second reservoir.

Among all the volumes for the MAS storage reservoir, the difference in efficiency is 8% compared to the largest and smallest reservoirs, and the need for external supply is only 14%, or 65 m3 per month.

For simulation 3, which covers a catchment area of 10,544 m^2, the results applied to the Analysis and Simulation Method and the Australian Practical Method can be seen in Table 10 and summarized in Figure 15.

Table 9. Reservoir sizing for simulation 3 with monthly rainfall data.

	Source method	Reservoir volume	Average monthly supply	Monthly Overflow Average	Efficiency
		m^3	m3. month-1	m3. month-1	%
Practical Method Australian	Rippl	2946	287	13,5	49%
	DSC	1457	314	41	44%
	MAS/MPA	1000	329	58	43%
	MAS/MPA Brazilian	500	361	91	34%
	MAS/MPA	455	366	96	33%
		250	389	119	30%
	German	104	411	142	28%
	English	102	412	142	28%
Method Analysis and Simulation	Rippl	2946	283	14	49%
	DSC	1457	307	43	45%
		1000	322	59	43%
	MAS/MPA	500	354	93	35%
	MAS/MPA Brazilian	455	359	98	34%
	MAS/MPA	250	382	122	31%
	German	104	405	144	28%
	English	102	405	144	28%

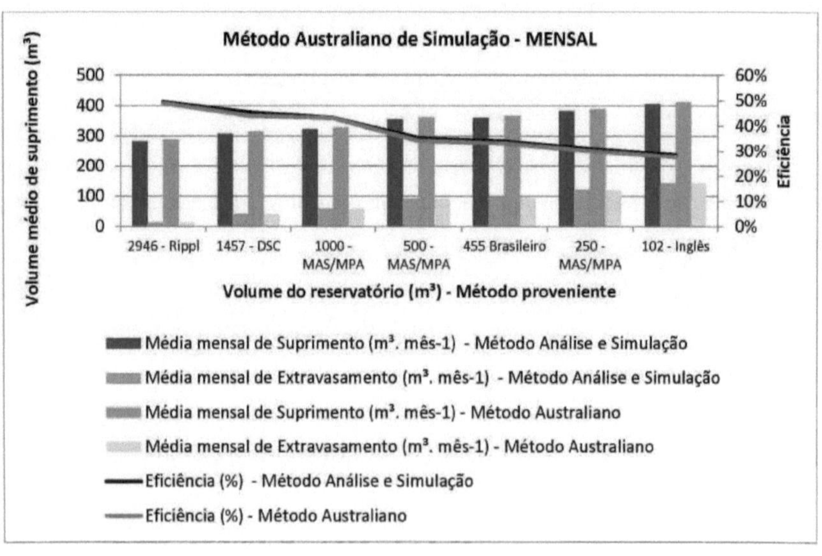

Figura 15. Reservoir sizing for simulation 3 with monthly rainfall data

Looking at the data from the third simulation and adopting random reservoir volumes to compare the best alternative, we opted for volumes of 2,946 m^3, 1,000 m^3 and 102 m^2. Applying the largest reservoir, the efficiency of meeting the water demand is approximately 50%, and it would still require an average of 283 m3 per month of water from an external source in order to carry out its activities in full.

If they chose to install a reservoir 3 times smaller, 1000 m3, the need for an external water supply would increase by 14%, resulting in 322 m3.month^{-1}. Even so, by installing a reservoir 29 times smaller, the water supply would increase by 43% (122 m3) compared to the first reservoir and 26% (83 m3) compared to the second.

Comparing the 3 simulations carried out, it is possible to determine that the larger the catchment area, the greater the volume of losses considered by the MPA and consequently the greater the difference in the need for an external water supply to carry out the company's activities, compared to the MAS.

Werneck (2006) comments that regardless of the method used, the calculation of reservoir volume is more accurate when daily consumption and rainfall data is used, with the aim of minimizing the risk of rainwater shortages in the system. To this end, under the same parameters applied in the simulations presented above (area, demand and fixed volume of the reservoir), the water balance of the reservoirs was carried out using daily rainfall and demand data.

For simulation 1, Table 11 and Figure 16 show the results of the sizing of the rainwater storage tank.

Table 10. Reservoir sizing for simulation 1 with daily rainfall data.

	Source method	Reservoir volume	Average monthly supply	Monthly Overflow Average	Efficiency
		m^3	m3. day-1	m3. day-1	%
Method Analysis and Simulation	Rippl	7785	22	0	43%
		1456	23	0	43%
	DSC	1000	23	0	43%
	MAS/MPA	290	25	3	36%
	Brazilian MAS/MPA	100	29	7	26%
	English	66	31	8	22%
	German	65	31	8	22%
Australian Practical Method	Rippl	7785	28	0	37%
		1456	28	0	37%
	DSC	1000	28	0	37%
	MAS/MPA	290	31	2	32%
	Brazilian MAS/MPA	100	35	6	22%
	English	66	36	8	20%
	German	65	36	8	20%

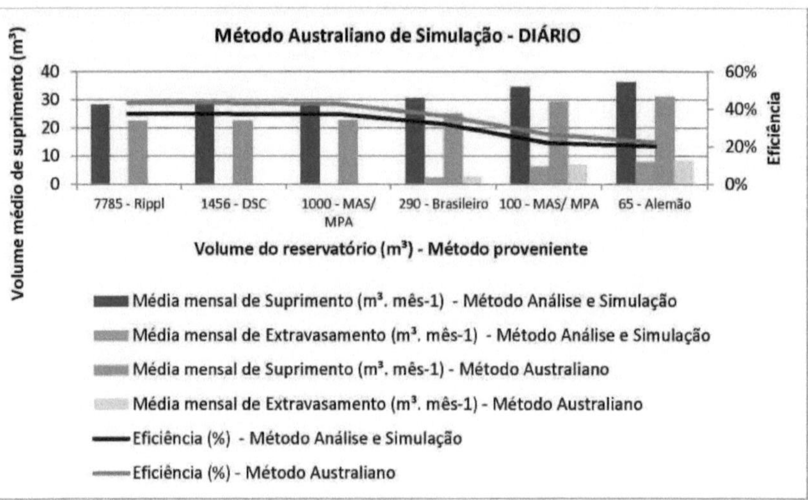

Figura 16. Reservoir sizing for simulation 1 with daily rainfall data

Daily rainfall data offers greater reliability for sizing and choosing the volume of the reservoir. For simulation 1, in order to guarantee the best water demand, adopting a volume of 1,000 m^3 corresponds to the same efficiency compared to a reservoir 8 times larger. However, adopting the 290 m3 reservoir would also meet a good part of the necessary water demand, with the help of approximately 31 m3 of water from an external source.

The sizing of the reservoir for the simulation of area 2 is shown in the Table and represented in Figure 17.

Table 11. Reservoir sizing for simulation 2 with daily rainfall data

	Source method	**Reservoir volume** m^3	**Average monthly supply** m3. day-1	**Monthly Overflow Average** m3. day-1	**Efficiency** %
Method Analysis and Simulation	Rippl	5522	16	0	59%
		1457	17	1	57%
		370	21	5	47%
	Brazilian DSC MAS/MPA	100	28	11	31%
	German	84	28	12	27%
	English	83	29	12	27%
	MAS/MPA	50	31	14	25%
Australian Practical Method	Rippl	5522	24	0	49%
		1457	24	0	48%
		370	28	4	40%
	Brazilian DSC MAS/MPA	100	34	10	25%
	German	84	35	11	24%
	English	83	35	11	24%
	MAS/MPA	50	37	13	17%

In this case, adopting the 1,457 m^3 reservoir, which is approximately 4 times smaller than the largest reservoir shown, would maintain efficiency at approximately 57%, requiring an average of 24 m3 per day of water from the external supply. However, adopting the 370 m3 reservoir would also meet a good part of the water demand and the size of the reservoir would be 4 times smaller.

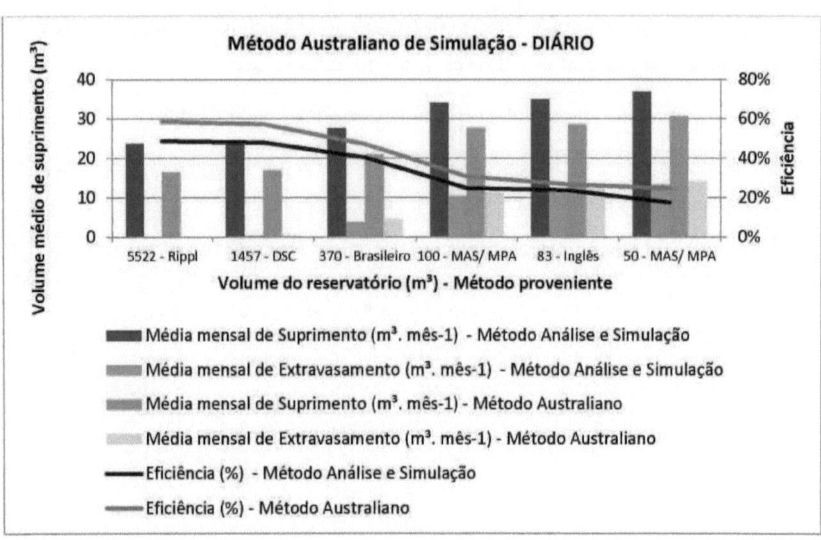

Figura 17. Reservoir sizing for simulation 2 with daily rainfall data

The sizing for simulation 3 with daily rainfall data can be seen in Table 13 and Figure 18, applied to the MAS and MPA methods.

Table 12. Reservoir sizing for simulation 3 with daily rainfall data.

	Source method	Reservoir volume	Average monthly supply	Average monthly overflow	Efficiency
		m^3	m3. day-1	m3. day-1	%
Analysis and Simulation Method	Rippl	2946	11	0	73%
	DSC	1457	12	2	70%
	MAS/MPA	1000	13	3	67%
	Brazilian	455	17	7	57%
	German	104	26	16	34%
	English	102	26	16	34%
	MAS/MPA	50	30	20	27%
Australian Practical Method	Rippl	2946	19	0	60%
	DSC	1457	20	1	58%
	MAS/MPA	1000	21	2	56%
	Brazilian	455	25	6	48%
	German	104	34	15	27%
	English	102	34	15	27%
	MAS/MPA	50	37	18	19%

For simulation 3, the reservoir with a capacity of 1,000 m^3 represents 67% of efficiency, requiring approximately 21 m3 of water per day from an external source. On the other hand, adopting a reservoir 2 times smaller would reduce efficiency by 10%, with a need for 25 m3 of water per day.

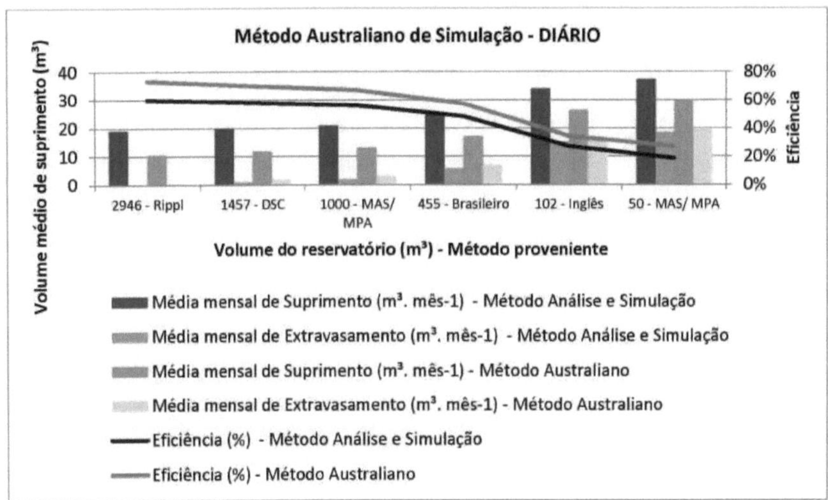

Figura 18. Reservoir sizing for simulation 3 with daily rainfall data

The efficiency of the system is an important parameter to be verified when sizing the collection system, however, this must be analyzed together with other factors such as spatial characteristics, investment and risk, in order to maintain sufficient service according to the needs and possibilities of the enterprise.

When comparing the efficiency resulting from the application of daily and monthly rainfall data for both methods (MAS and MPA), there is a significant difference. To obtain monthly data, the rainfall for each day of the month is added up and then the need for water supply or overflow from the reservoir is analyzed.

Efficiency, when using monthly data, only considers whether or not the monthly water demand has been met. By analyzing daily data, it is possible to see the days when the system was in deficit, i.e. when there were too many overflows or when the reservoir was not filled.

In this respect, the results of the daily sizing can be seen in Figures 9, 10 and 11, which show the daily distribution of the volume of rain that can be used, being compatible with the minimum reservoir volumes of 1,000 m^3, 1,457m^3 and 1,000 m$^{(3)}$for simulations 1, 2 and 3 respectively, with the aim of capturing the largest volume of rain and storing it for future periods of drought.

4.2 Pumping stored water

The purpose of sizing the booster pump is to measure the average electricity consumption that will be spent on the new collection system. To this end, a number of factors were taken into account for the sizing, mainly the actual dimensions of the project for its installation.

Considering that the water storage reservoir will be supported by the ground and the motor pump chosen is classified as "multi-stage 5 submersible", model VN-5312, designed to operate in water, with a maximum submergence of 20 meters and is suitable for transporting water over long distances and also for collecting rainwater.

With these characteristics, the pump will be submerged in the reservoir and will discharge water to a height of 5.5 m in a straight line, and then, by inserting special parts, it will surround the Company for a length of 151 m, when it will descend to a height of 6 meters until it reaches the use reservoir, which is arranged underground (Appendix D).

The company currently has two underground tanks next to the vehicle washing machines, one with a capacity of 17m3 and the other 30m3. These tanks are already pumped to the washing machine. In view of this, the system's savings in relation to electricity will not be verified, only the sizing of the pump for the first rainwater storage tank, which is being sized in this work, will be carried out.

The discharge flow is 3 $m3.h^{-1}$, and some special parts were considered for the water discharge: 3 90° bends, 1 gate valve, 1 check valve and 1 concentric extension. By calculating the pressure drop of both the parts and the discharge, the total head was 5.62 m.

As recommended by SRHT (2007), it is important that the manometric height is slightly greater than that measured in the field. A gap of 0.5 m was adopted, giving a total head of 6.12 meters.

The manufacturer checked was Scheneider motobombas (2016), and the characteristics of the pump required were a rotor diameter of 97 mm, a maximum head of 34 m.c.a (meters of water column) and a power of 1.2 hp.

Given the power of the pump, it needs to work 24 hours a day, 30 days a month and, considering the consumption of R$1.05 $R\$.kwh^{-1}$, this results in an average electricity cost of approximately R$665.28 per month.

4.3 Economic viability of the system

The analysis of the results regarding the economic viability of implementing the rainwater harvesting system can be seen in Appendices G, H and I for simulations 1, 2 and 3 respectively.

The results are presented in terms of consumption, cost and savings.

Consumption is expressed as the volume of water needed to meet the demand for water to carry out the company's activities, both before the abstraction system is installed and after it is installed, by applying the external water supply volume data for each reservoir volume applied.

The cost is measured through water consumption to meet the company's water demand, for the two situations: before and after the installation of the collection system. The water and sewage tariff applied by SANEP in the municipality of Pelotas was taken into account, and, according to the City Hall, sewage collection and treatment for the site of the project is 80%.

Finally, savings are measured by the difference between the situations before and after the installation of the rainwater harvesting system at the company, according to consumption characteristics.

The data was presented only for the monthly simulations carried out, since it was not possible to calculate the daily financial water consumption according to the SANEP billing criteria for the municipality of Pelotas.

Figure 19 shows simulation 1, and it can be seen that the difference in financial savings when choosing one of the methods (MPA or MAS) for sizing the rainwater storage reservoir is very low, being close to 1%, representing no significant difference in choice.

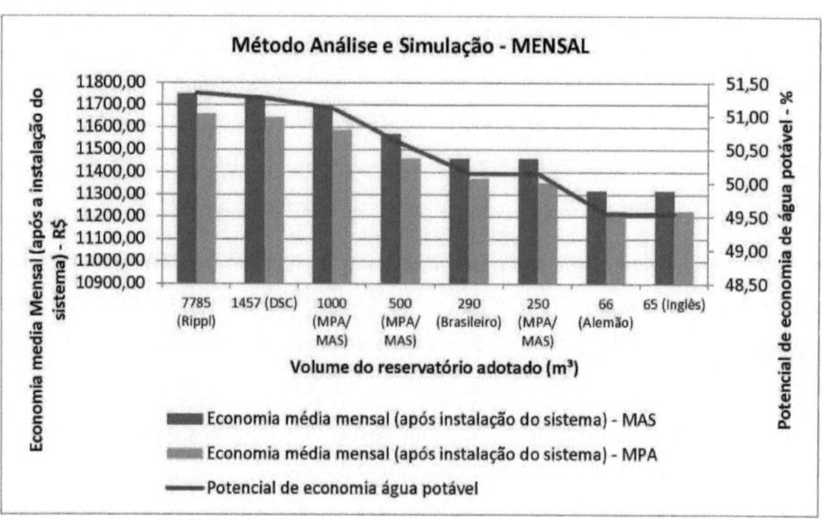

Figure 19. Average monthly water savings for simulation 1.

It is also possible to see that the potential savings in drinking water are approximately 49%, i.e. around 683,000 liters of water per month for the largest reservoir and 660,000 liters of water for the smallest reservoir.

As far as the financial aspect is concerned, there is an average monthly reduction of up to R$11,316.20 for the smallest water storage reservoir and R$11,751.30 for the largest reservoir, representing a difference in savings of approximately 4% between the two reservoirs.

Figure 20 shows the data for simulation 2. It can be seen that with the possibility of expanding the water catchment area, the potential for saving drinking water reaches approximately 65% by adopting the largest reservoir and 60% by adopting the smallest, representing an average of 870,000 liters of water per month for the first option and 805,000 liters of water for the second.

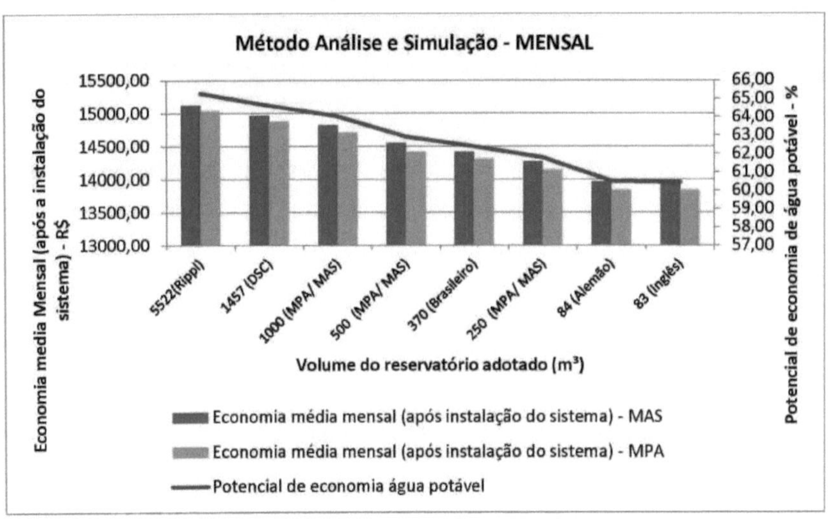

Figure 20. Average monthly water savings for simulation 2.

The difference in the average reduction in measurable costs after applying the rainwater harvesting system represents around 50% of the costs for each reservoir dimensioned.

As for water consumption, the financial reduction is on average 65% if the largest reservoir is adopted and 60% if the smallest is adopted, representing R$15,123.50 for the largest reservoir and R$13,945.00 for the smallest.

Figure 21 shows the data for the third simulation of the catchment area.

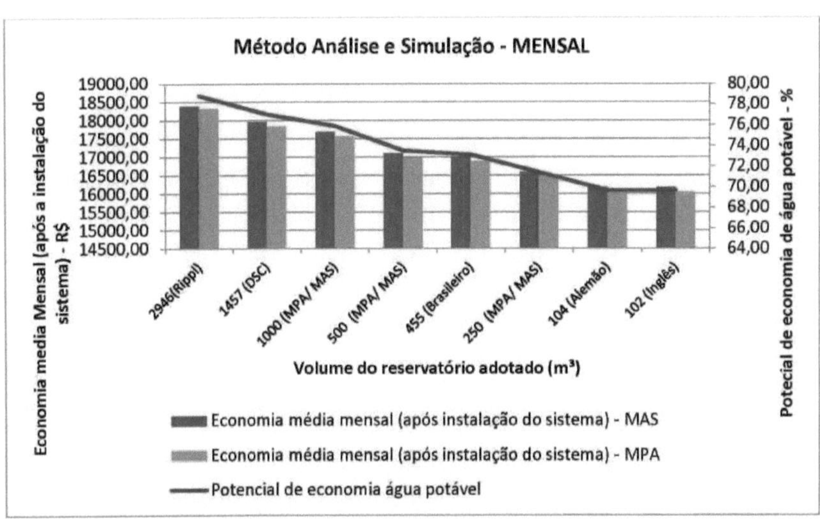

Figure 21. Average monthly water savings for simulation 3.

For simulation 3, it can be seen that, in financial terms, the average saving in water alone is around 80% to 70% after installing the system.

If the largest reservoir is used, the savings in potable water from the urban supply represent an average of 1,050,000 liters of water per month and, for the smallest reservoir, approximately 928,000 liters of water, i.e. approximately 80% for the first case and 70% for the second.

The average cost reduction between the largest and smallest reservoirs adopted is approximately 13%, with R$18,386.90 for the largest and R$16,175.00 for the smallest.

For the 3 simulations carried out, the variability in the choice of method (MAS or MPA) was not significant.

The larger the catchment area, the greater the savings in the volume of water from the external supply needed to meet demand, and the less it will cost. However, even adopting the smallest catchment area and the smallest reservoir volume for storing water (65 m^3), the saving is approximately R$11,316.20 per month, which means R$377.00 per day.

It is important to note that the data presented for all the cases studied are monthly averages, and may vary depending on the intensity of rainfall in the municipality. However, it is possible to observe that, for all cases, rainwater harvesting was satisfactory both in financial terms, through the reduction of expenses related to water consumption, and in relation to the environment, through the conservation of water for more noble uses.

In addition, in this work no budgets were made for the construction and implementation of the system, in order to guarantee the viability of the investment and its payback period for the company.

Looking at Figures 19, 20 and 21, it can be seen that none of the simulations met the 100% water saving requirement, since this is dependent on natural factors, such as rainfall, even though this is satisfactory for the region. To this end, it is necessary to connect to the municipality's external supply network, so as not to compromise the operation of the project's activities.

The water collection reservoir has not been designed to dispose of the first rainfall, which is considered adequate for the vehicle washing system. However, it is hoped that filters will be installed in the pipes to prevent leaves and debris from entering the reservoir, as well as the addition of chlorine to inhibit the development of bacteria and prevent the proliferation of diseases that could be transmitted by it.

5. CONCLUSION

According to the simulations of different catchment areas and the volume of rainwater that can be retained by the system, it can be concluded that the municipality of Pelotas has potential for rainwater catchment, since the annual rainfall is regular and evenly distributed and the municipality does not suffer from long periods of drought.

An important factor for the viability of using rainwater in the enterprise is the size of the catchment area that can be used. However, due to the high volume of water needed to carry out daily activities, this factor becomes insufficient, as large volumes of reservoirs are needed to store the water, in order to increase the efficiency of the system and reduce costs with external water supplies to carry out the company's activities.

For this reason, efficiency will not always be a decisive parameter for implementing the system, since, as seen in the simulations carried out, greater efficiency requires larger volumes of reservoirs, which makes implementation unfeasible due to the need for space.

The sizing methods for rainwater storage reservoirs presented in NBR 15.527 (2007) result in different reservoir volumes due to their diverse calculation methodology. However, when these volumes are applied to the Analysis and Simulation Method and/or the Australian Practical Method, they make it possible to analyze the water balance that will take place inside the reservoir and check which is the best alternative for each case, taking into account technical, economic and environmental factors.

The interpretation of the data resulting from the water balance is an important factor to be analyzed together, verifying the needs that the system will demand, in order to size the best reservoir volume, considering costs, investment needs and the spatial issues available for installation.

The Analysis and Simulation Method and the Australian Practical Method proved to be very efficient in the results presented, as well as being easy to manipulate, and the use of daily rainfall data provides greater confidence in decision-making due to the possibility of checking the water balance and efficiency of the system on a daily basis, its needs and deficiencies.

Analyzing the simulations carried out, we recommend using the largest area (10,544 m^2) and applying a 455 m^3 reservoir, with the possible average need for 12,000 liters of water per day from the external supply, representing an efficiency of 34% of the entire system.

This reservoir represents a saving of 73% of water, which contributes to a reduction in the use of surface water resources and a reduction in water supply costs of approximately R$17,009.00 per month. This is the difference between the situation before the rainwater harvesting system was installed and after it was fully operational.

However, even when using the smallest reservoir (102 m3) for the largest catchment area, it proved to be viable due to the reduction in water consumption and also the reduction in costs for the project, which amounted to approximately R$16,175.00 per month.

It is important to emphasize that some practices should be rethought by the municipality's public administration, by lowering the IPTU rate charged to homes, businesses and enterprises that adopt sustainable measures, such as the installation of rainwater harvesting systems, in order to motivate and engage the community in the rational use of natural resources.

The implementation of these actions would promote greater participation by the population, which would benefit from lower water consumption costs, as well as conserving water and putting it to better uses.

It is recommended that the quality of the rainwater be assessed and verified in order to analyze the need to dispose of the first millimetres of it, since the current project does not provide for this.

In addition, for future projects, it is proposed to analyze the water resulting from washing the vehicles, in order to verify the possibility and feasibility of inserting an effluent treatment system, in order to reuse it again in the company's processes, acting as a closed cycle.

REFERENCE

ABCMAC. Associação Brasileira de Captaçâo e Manejo de Agua de Chuva. Available at:< http://www.abcmac.org.br>. Accessed on: Feb. 3, 2016.

AMORIM, Simar Vieira de; PEREIRA, Daniel José de Andrade. Comparative study of sizing methods for rainwater harvesting reservoirs. In: Encontro Nacional de Tecnologia do Ambiente Construido, 12., Anais... Fortaleza, 2008.

ANA - National Water Agency. Available at:< http://www2.ana.gov.br/Paginas/default.aspx>. Accessed on: March 13, 2016.

ANA, FIESP & SINCUSCON-SP. Conservation and reuse of water in buildings: Sâo Paulo: Prol Editora Gràfica. 2005. 152 p.

ANNECCHINI, K. P. V. Using rainwater for non-potable purposes in the city of Vitória (ES). 150f. Dissertation (Master's in Environmental Engineering) - Technology Center, Federal University of Espirito Santo, Vitória, 2005.

Statistical Yearbook of Brazil - IBGE. Volume 74. 2014. Available at: <http://biblioteca.ibge.gov.br/visualizacao/periodicos/20/aeb_2014.pdf>. Accessed on: Feb. 3, 2016.

Guarani Aquifer. Departamento Autònomo de Agua e Esgoto (DAAE). Available at: <http://www.daaeararaquara.com.br/guarani.htm>. Accessed on: February 3, 2016.

ASA - ARTICULAÇÃO NO SEMI-ARIDO BRASILEIRO. Available at:< http://www.asabrasil.org.br/> Accessed on: Feb. 15, 2016.

BRANCO, C. C. de O.. Using rainwater in a car wash. Monograph (Graduation in Environmental Engineering) - Faculdade Dinâmica de Cataratas - (UDC), Foz do Iguaçu, Paranà, 2010.

BRAZIL. Commission on Sustainable Development Policies and Agenda 21. In: *Brazilian Agenda 21*. Brasilia, 2002. Available at:

<http://www.mma.gov.br/responsabilidade-socioambiental/agenda-21/agenda-21-Brazilian>. Accessed on: Feb. 15, 2016.

BRAZIL. Decree No. 24.643, of July 10, 1943 - Institutes the Water Code. Rio de Janeiro, RJ, July 10, 1934.

BRAZIL. Federal Law 9433 of January 8, 1997. Institutes the National Water Resources Policy. Brasilia, DF, January 8, 1997.

BRAZIL. Federal Law No. 11.445, of January 5, 2007. Institutes the National Sanitation Policy. Brasilia, DF, January 2007.

BRAZIL. Bill No. 7818 of 2014. Establishes the National Policy for the Collection, Storage and Use of Rainwater and defines general rules for its promotion. Brasilia, DF, 2014.

BRITTO, F. P. et al. Spatial and temporal variability of rainfall in Rio Grande do Sul: Influence of the el nino southern oscillation. Brazilian Journal of Climatology. ISSN: 1980-055X, 2008.

PELOTAS CITY COUNCIL. Municipal Law No. 6.294, of December 2, 2015. Provides for changes to the system for charging for water supply and wastewater collection and treatment by the Autonomous Sanitation Service of Pelotas - SANEP. Pelotas, RS, December 2015.

PORTO ALEGRE CITY COUNCIL. Municipal Law n. 10.506, of August 5, 2008. Establishes the Water Conservation, Rational Use and Reuse Program.

CARVALHO, G. F.; OLIVEIRA, S. C & MORUZZI, R. B. Calculation of the reservoir volume of rainwater harvesting systems: comparison between methods for application in single-family homes. In: SIMPÒSIO NACIONAL DE SISTEMAS PREDIAIS, 10, 2007, Sâo Carlos. Proceedings... Sâo Carlos: UFSCar, 2007. 1 CD-ROM.

CHRISTOFIDIS, D. Agua, ética, segurança alimentar e sustentabilidade ambiental. BAHIA ANALISE & DADOS, Salvador, v. 13, n. especial, p. 371-382, 2003.

CIRRA - Centro Internacional de Referência em Reùso de Agua; FCTH - Fundaçâo Centro Tecnològico de Hidraulica; DTC Engenharia. Water Conservation and Reuse: Guidance Manual for the Industrial Sector. FIESP/CIESP, São Paulo, 2004.

CILENTO, F. C. Solutions for the use of rainwater in existing buildings through the development of disposal and storage reservoirs. Dissertation (Master's Degree in Environmental Technology) - Instituto de Pesquisas Tecnológicas do Estado de Sâo Paulo, Sâo Paulo, 2009.

COHIM, E. et al. Direct rainwater harvesting in the urban environment for non-potable uses. In: Brazilian Congress of Sanitary and Environmental Engineering, 24th, 2007, Belo Horizonte. Proceedings... Belo Horizonte: ABES, 2007. p. 13.

COHIM, Eduardo; GARCIA, Ana; KIPERSTOK, Asher. Rainwater harvesting and utilization: reservoir sizing. IX Symposium on Water Resources of the Northeast. Salvador, BA. 2008.

Conjuncture of Water Resources in Brazil. ANA/ MMA. Brasilia, 2013.

DEVKOTA, J., Schlachter, H., & Apul, D. Life cycle based evaluation of harvested rainwater use in toilets and for irrigation. *Journal of Cleaner Production*, 311-321. ISSN: 0959-6526. 2015.

DORNELLES, F. Utilization of rainwater in the urban environment and its effect on rainwater drainage. PhD thesis - Institute of Hydraulic Research (IPH) - UFRGS. Porto Alegre, 2012.

DORNELLES, Fernando; TASSI, Rutinéia; GOLDENFUM, Joel A. Evaluation of sizing techniques for rainwater harvesting reservoirs. Revista Brasileira de Recursos Hidricos, Volume 15 n. 2 Apr/ Jun. 2010, 5968.

EAP - Pelotas Agroclimatological Station. Available at: <http://agromet.cpact.embrapa.br/estacao/estacao.html>. Accessed on April 10, 2016.

FREITAS, V. P. Waters - Legal and environmental aspects. Curitiba: Juruà, 2000.

GARCEZ, L. N.; ALVAREZ, G. A. Hidrologia. 2 ed., Sâo Paulo: Edgard Blucher, 1988.

GHISI, E. The Influence of Rainfall, Catchment Area, Number of Residents and Potable and Rainwater Demand on the Dimensioning of Reservoirs for the Use of Rainwater in Single-Family Residences. Monograph presented to the Department of Civil Engineering of the Federal University of Santa Catarina as part of the requirements for participation in the Public Tender of Notice No. 026/DDPP/2006.Florianópolis, 2006.

GNADLINGER, J. Rainwater Harvesting in Rural Areas. International Association of Rainwater Harvesting Systems. 2nd World Water Forum, The Netherlands, 2000. Available at:< http://www.irpaa.org.br/colheita/indexb.htm> Accessed on: Feb. 20, 2016.

GROUP RAINDROPS. Harnessing Rainwater. Editora Organic Trading, 1ª Ediçâo, Curitiba, 2002.

GUEDES, R. L. Large-scale conditions associated with mesoscale convective systems over the Central Region of South America. Sâo Paulo, 1985.

Master's Degree Dissertation - Astronomical and Geophysical Institute, University of Sao Paulo.

HAGEMANN, S. E. Evaluation of the quality of rainwater and the feasibility of its collection and use. Dissertation - Postgraduate Program in Civil Engineering Area of Concentration in Water Resources and Environmental Sanitation,. UFSM - RS. Santa Maria, 2009.

IBGE - Brazilian Institute of Geography and Statistics. Demographic census. Available at: <http://www.ibge.gov.br/home/estatistica/populacao/censo2010/default.shtm>. Accessed on: Feb. 3, 2016.

INMET - National Institute of Meteorology. Available at: <http://www.inmet.gov.br/portal/index.php?r=home2/index>. Accessed on: February 29, 2016.

KINKER, R. S. Proposal for implementing rainwater harvesting in social housing developments: a case study. Dissertation (Master's Degree in Housing: Planning and Technology) - Technological Research Institute of the State of Sao Paulo. Sao Paulo, 2009.

KOBIYAMA, M.; TSUYOSHI, U. ANJOS AFONSO, M. (organizers of the translation). Making use of rainwater - Ed. Organic Trading, 1st edition, 196p - Curitiba/ PR - Brazil, 2002.

LOBATO, M. B. System for prioritizing water conservation actions in buildings using the Electre III method. 283 f. Dissertation (Master's Degree in Civil Construction) - Federal University of Paranà, Curitiba, 2005.

MAESTRI, R. S. Cost-Benefit Analysis of Rainwater Utilization in Florianópolis. Course Conclusion Work in Sanitary and Environmental Engineering: UFSC. Florianópolis, 2003.

MANO, R. S.; SCHMITT, C. M. Residential Rainwater Harvesting for Non-Potable Purposes in Porto Alegre: Basic Aspects of the Technical Feasibility and Benefits of the System. CLACS' 04 - I Conferencia Latino-Americana de Construçâo Sustentâvel and ENTAC 04, - 10° Encontro Nacional de Tecnologia do Ambiente Construido, Sâo Paulo - SP, Anais....CD Rom, 2004.

MARINOSKI, A. K. Using rainwater for non-potable purposes in an educational institution: A case study in Florianópolis - SC. 2007. 107 f. Course Conclusion Paper (Graduation in Civil Engineering) - Technology Center, Department of Civil Engineering, Federal University of Santa Catarina, Florianópolis, 2007.

MAY, S. Study of the Feasibility of Using Rainwater for Non-Potable Consumption in Buildings. Dissertation (Master's in Engineering). Postgraduate Course in Civil Construction Engineering. Polytechnic School. University of Sâo Paulo, Sâo Paulo, 2004.

MAY S.; PRADO R. T. A. Study of Rainwater Quality for Non-Potable Consumption in Buildings. CLACS' 04 - I Latin American Conference on Sustainable Construction and ENTAC 04, - 10th National Meeting on Built Environment Technology, Sâo Paulo - SP. Annals....CD Rom, 2004.

MIERZWA, J. C. et al. Rainwater: reservoir calculation method and concepts for proper use. REGA - Vol. 4, n° 1, p. 29-37, Jan./Jun. 2007.

MINISTRY OF HEALTH. Ordinance No. 518 of March 25, 2004. Available at:

<http://www.aeap.org.br/doc/portaria_518_de_25_de_marco_2004.pdf>. Accessed on: Feb. 25, 2016.

MINISTRY OF THE ENVIRONMENT. SUSTAINABLE CONSUMPTION: Education manual. Brasilia: Consumers International/ MMA/ MEC/IDEC, 2005. 160 p.

MORELLI, E. B. Water Reuse in Vehicle Washing. Dissertation (Master's Degree in Engineering). Postgraduate Course in Civil Construction Engineering. Polytechnic School. University of São Paulo, 2005.

MORUZZI, Rodrigo Braga; OLIVEIRA, Samuel Conceiçâo de. Application of a computer program in the sizing of reservoir volume for the rainwater harvesting system in the city of Ponta Grossa, PR. Journal of Engineering and Technology. V. 2, N° 1, April/2010.

NBR 15527: Rainwater - Use on roofs in urban areas for non-potable purposes - Requirements. 2007.

NBR 10844: Instalaçôes prediais de águas pluviais , 1989.

NBR 5.626: Cold water building installation, 1998.

NBR 12.214: Design of a water pumping system, 1992.

NOLDE, E. Possibilities of rainwater utilization in densely populated areas including precipitation runoffs from traffic surfaces . *Desalination* , 1-11. 2007.

OECD, Organization for Economic Cooperation and Development, Report: Pricing Water Resources and Water and Sanitation Services, Paris - France, 2010.

PHILIPPI, L.S. et al. Making use of rainwater. In: GONÇALVES, R.F. (Org.). Rational use of water in buildings. Rio de Janeiro: ABES - PROSAB, 2006. chap. 3, p. 73-152.

REBOUÇAS, Aldo Rebouças. Intelligent Use of Water. São Paulo. Escrituras, 2004.

SANTOS, C., & Taveira-Pinto, F. Analysis of different criteria to size rainwater storage tanks using detailed methods. *Resources, Conservation and Recycling*, 1-6. ISSN: 0921-3449. 2013.

SAUTCHÛK, C. A. Formulation of guidelines for the implementation of water conservation programs in buildings. Dissertation (Master's in Engineering) - Polytechnic School of the University of Sâo Paulo, Sâo Paulo, 2004.

SBRT. Brazilian Technical Response Service. Available at: http://sbrt.ibict.br. Accessed on: June 3, 16.

SCHNEIDER MOTOR PUMP. Available at:

<http://www.schneider.ind.br/produtos/motobombas-submers%C3%ADveis/submers%C3%ADveis/vn/>. Accessed on: June 9, 2016.

SEEGER, L. M. K.; SARI, V.; PAIVA, E. M. C. D. Comparative analysis of the use of rainwater for washing vehicles in two cities in the South and Midwest regions. In: SIMPÓSIO BRASILEIRO DE RECURSOS HiDRICOS, 17., 2007, São Paulo. Proceedings... São Paulo: [s.n.], 2007. 1-13.

SEGALA, M. Water: scarcity in abundance. Sustainable Planet. Available at: *<http://planetasustentavel.abril.com.br/noticia/ambiente/populacao-falta-agua- recursos-hidricos-graves-problemas-economicos-politicos-723513.shtml>.*

Accessed on: April 10, 2016.

SENRA, J.B; BRONZATTO, L.A; VENDRUSCOLO, S. Rainwater Harvesting in the National Water Resources Plan. Rainwater: research, policies and sustainable development. In: *Proceedings* of the VI Brazilian Symposium on Rainwater Harvesting and Storage, Belo Horizonte, 2007.

SILVA, Eduardo Rosa da. Using rainwater for non-potable consumption in gas stations. Canoas. 75 p. Course Conclusion Paper, Civil Engineering, ULBRA, 2007.

SOARES, D. A. F. et al. Sizing a rainwater reservoir to assist toillet flushing. In: CIB W62 Seminar, Rio de Janeiro. Proceedings. CIB W62 Seminar, Rio de Janeiro, v. 1, p.D11 - 1D1 - 12, 2000.

TOMAZ, P. Forecasting water consumption. São Paulo: Navegar Editora, 2000.

TOMAZ, P. Harnessing Rainwater - For Urban Areas and Non-Potable Purposes. Navegar Editora, São Paulo, 2003.

TORDO, O. C. Characterization and evaluation of the use of rainwater for drinking purposes. Dissertation (Master's Degree) - Center for Technological Sciences and Postgraduate Program in Environmental Engineering, Regional University of Blumenau. Blumenau, 2004.

TUCCI, C. E. M. Hidrologia: Ciência e Aplicaçâo: 2. ed. Porto Alegre: Ed. Universidade/UFRGS: ABRH, 2001. 943 p.

TUCCI, C. E. M., Urban Flooding. Collection ABRH V, 11, Porto Alegre/ RS - Brazil, 2007.

UNEP - United Nations Environment Programme. Global Environment Outlook 3: past, present and future perspectives. London: Earthscan, 2002.

VILLIERS, M. Agua: How the use of this precious natural resource could lead to the most serious crisis of the 21st century. Rio de Janeiro: Ediouro, 2002.

WATERFALL, P.H.. Harvesting Rainwater for Landscape Use. University of. Arizona Cooperative (2002). Available at: < http://ag.arizona.edu/pubs/water/az1052/ >. Accessed on: Feb. 10, 2016.

WERNECK , G A. M.; Rainwater utilization systems in buildings: a case study of the application in a school in Barra do Pirai. Dissertation (master's) - Postgraduate Program in Architecture - UFRJ/ FAU/ PROARQ Rio de Janeiro: UFRJ/ FAU, 2006.

WORLD HEALTH ORGANIZATION (WHO). Health in water resources development (2006). Available at:< http://www.who.int/docstore/water_sanitation _health /vector/water_resources.htm>. Accessed on: March 24, 2016.

APPENDICES

Appendix A - Sketch of the layout of the Expresso Embaixador company - Simulation area 1

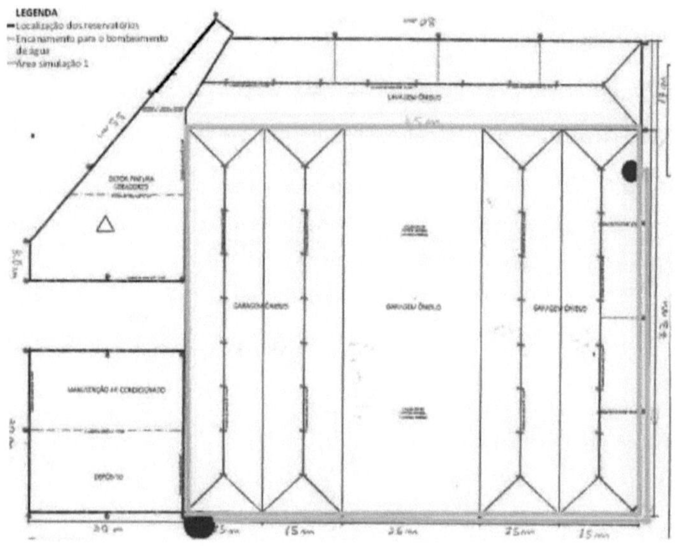

Appendix B - Sketch of the layout of the Expresso Embaixador Company - Simulation area 2

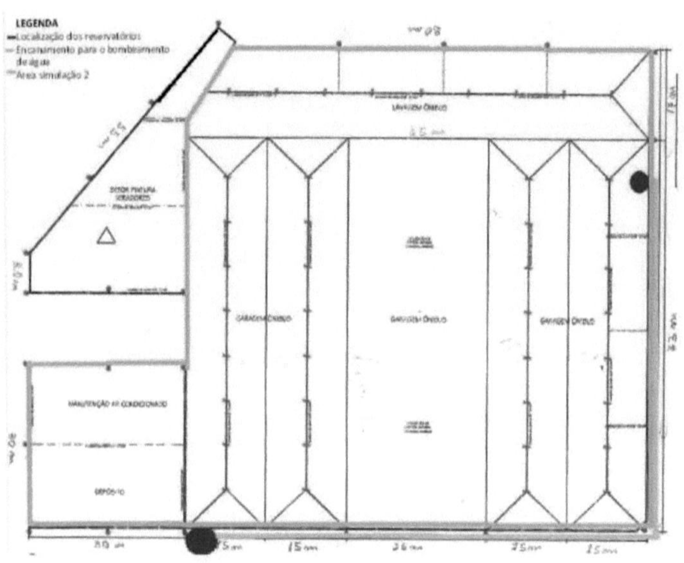

Appendix C - Sketch of the layout of the Expresso Embaixador company - Simulation area 3

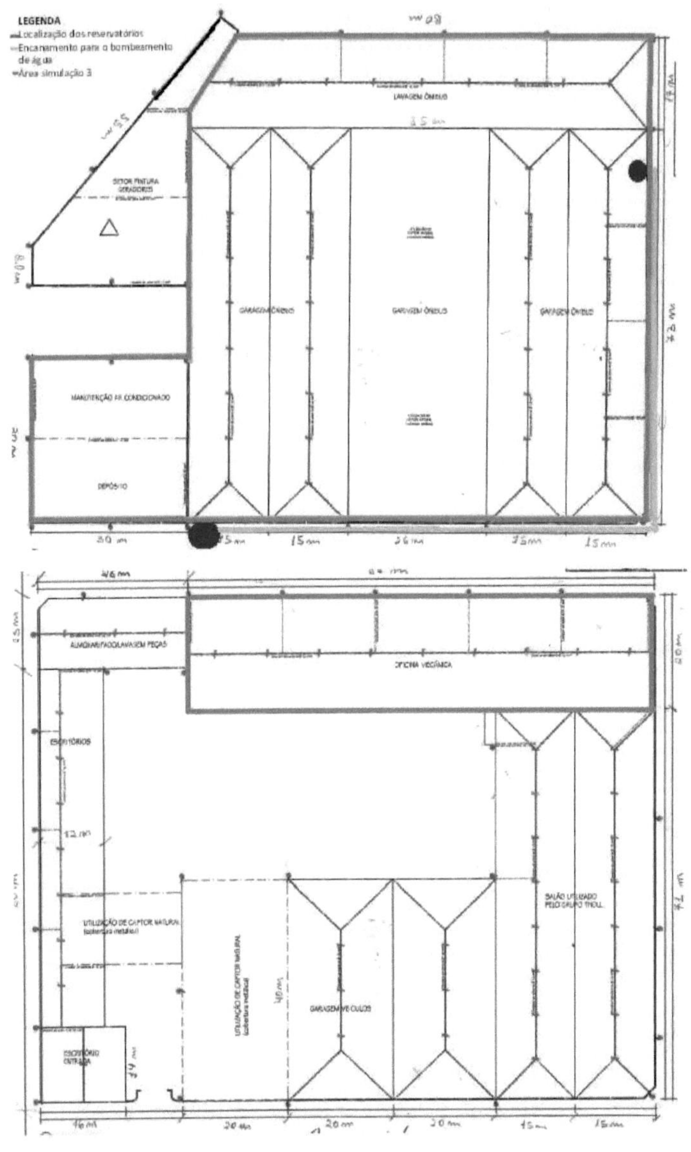

Appendix D - Sketch of the Expresso Embaixador plant - Location of the storage reservoir and pumping system

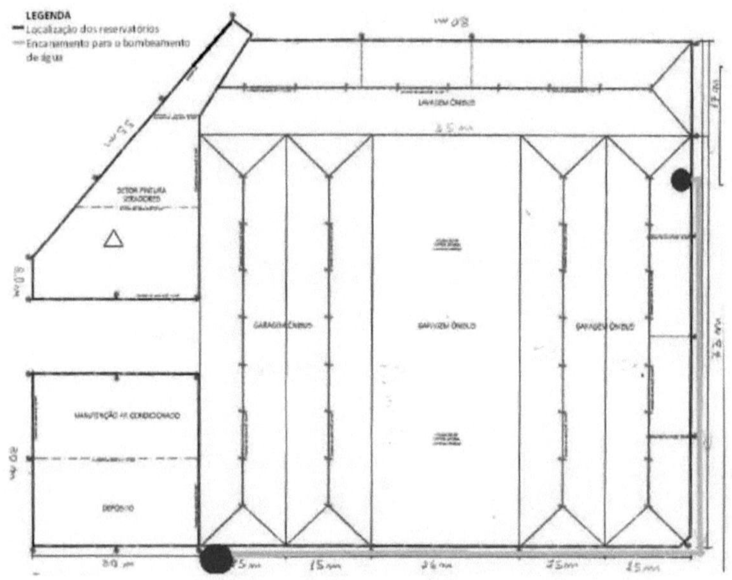

Appendix E - Average monthly rainfall data for the municipality of Pelotas from 1980 to 2015.

Months	Average monthly rainfall (mm)
January	114,59
February	163,42
March	110,09
April	115,64
May	116,74
June	115,35
July	124,18
August	113,28
September	131,94
October	108,99
November	113,31
December	109,23

Appendix F - Average annual rainfall data, consecutive days without rain and months in which rainfall exceeded 100 mm in the municipality of Pelotas from 1980 to 2015.

Year	Average annual rainfall (mm)	Months with precipitation< 100mm	Consecutive days without precipitation
1980	129,6	5	25
1981	111,8	6	27
1982	111,8	4	24
1983	128,6	7	22
1984	141,2	4	22
1985	98,2	7	27
1986	138,0	3	16
1987	149,5	5	17
1988	74,2	9	16
1989	68,6	10	34
1990	125,2	5	28
1991	111,1	5	21
1992	119,5	6	15
1993	115,8	5	16
1994	105,4	7	18
1995	135,9	5	18
1996	82,6	7	32
1997	157,8	4	24
1998	154,9	2	19
1999	93,2	8	20
2000	124,3	4	26
2001	156,3	3	14
2002	192,9	0	15
2003	121,7	6	27
2004	114,1	9	27
2005	90,7	9	36
2006	90,3	7	25
2007	123,2	5	33
2008	97,3	7	30
2009	142,3	7	38
2010	110,4	6	27
2011	91,4	7	22
2012	93,5	7	26
2013	105,7	7	19
2014	150,7	3	18
2015	152,7	3	13

Appendix G - Consumption ratio, costs and savings for the rainwater harvesting system, applied in simulation 1.

		CONSUMPTION		COST			ECONOMY	
Method	Volume reservoir	Consumption average monthly drinking water consumption (before system installation)	New average monthly drinking water consumption (after system installation)	Cost of drinking water (before the system is installed)	New cost average monthly drinking water (after system installation)	Cost system operation (electricity)	Potential saving drinking water	Economy monthly average (after system installation)
Unit	m³	m³.mès'¹	m³.month'¹	R$.month'¹	R$.month'¹	R$.month'¹	%	R$. month'¹
Australian monthly	7785	1333,86	654,00	24215,31	11889,37	665,28	50,97	11660,66
	1457	1333,86	655,00	24215,31	11907,50	665,28	50,89	11642,53
	1000	1333,86	658,00	24215,31	11961,89	665,28	50,67	11588,14
	500	1333,86	665,00	24215,31	12088,80	665,28	50,14	11461,23
	290	1333,86	670,00	24215,31	12179,45	665,28	49,77	11370,58
	250	1333,86	671,00	24215,31	12197,58	665,28	49,70	11352,45
	66	1333,86	678,00	24215,31	12324,49	665,28	49,17	11225,54
	65	1333,86	678,00	24215,31	12324,49	665,28	49,17	11225,54
Monthly simulation	7785	1333,86	649,00	24215,31	11798,72	665,28	51,34	11751,31
	1457	1333,86	650,00	24215,31	11816,85	665,28	51,27	11733,18
	1000	1333,86	652,00	24215,31	11853,11	665,28	51,12	11696,92
	500	1333,86	659,00	24215,31	11980,02	665,28	50,59	11570,01
	290	1333,86	665,00	24215,31	12088,80	665,28	50,14	11461,23
	250	1333,86	665,00	24215,31	12088,80	665,28	50,14	11461,23
	66	1333,86	673,00	24215,31	12233,84	665,28	49,55	11316,19
	65	1333,86	673,00	24215,31	12233,84	665,28	49,55	11316,19

Appendix H - Consumption ratio, costs and savings for the rainwater harvesting system, applied in simulation 2.

		CONSUMPTION		COST			ECONOMY	
Method	Volume reservoir	Consumption average monthly drinking water consumption (before system installation)	New average monthly drinking water consumption (after system installation)	Cost of drinking water (before the system is installed)	New cost average monthly drinking water (after system installation)	Cost system operation (electricity)	Potential saving drinking water	Economy monthly average (after system installation)
Unit	m³	m³.mès'¹	m³.month'¹	R$.month'¹	R$.month'¹	R$.month'¹	%	R$. month'¹
Australian monthly	5522	1333,86	468,00	24215,31	8517,19	665,28	64,91	15032,84
	1457	1333,86	477,00	24215,31	8680,36	665,28	64,24	14869,67
	1000	1333,86	486,00	24215,31	8843,53	665,28	63,56	14706,50
	500	1333,86	502,00	24215,31	9133,61	665,28	62,36	14416,42
	370	1333,86	508,00	24215,31	9242,39	665,28	61,92	14307,64
	250	1333,86	517,00	24215,31	9405,56	665,28	61,24	14144,47
	84	1333,86	534,00	24215,31	9713,77	665,28	59,97	13836,26
	83	1333,86	534,00	24215,31	9713,77	665,28	59,97	13836,26
Monthly simulation	5522	1333,86	463,00	24215,31	8426,54	665,28	65,29	15123,49
	1457	1333,86	472,00	24215,31	8589,71	665,28	64,61	14960,32
	1000	1333,86	480,00	24215,31	8734,75	665,28	64,01	14815,28
	500	1333,86	495,00	24215,31	9006,70	665,28	62,89	14543,33
	370	1333,86	502,00	24215,31	9133,61	665,28	62,36	14416,42
	250	1333,86	510,00	24215,31	9278,65	665,28	61,77	14271,38
	84	1333,86	527,00	24215,31	9586,86	665,28	60,49	13963,17
	83	1333,86	528,00	24215,31	9604,99	665,28	60,42	13945,04

Appendix I - Consumption, costs and savings for the rainwater harvesting system, applied in simulation 3.

		CONSUMPTION		COST		ECONOMY		
Method	Volume reservoir	Consumption average monthly drinking water consumption (before system installation)	New average monthly drinking water consumption (after system installation)	Cost of drinking water (before the system is installed)	New cost average monthly drinking water (after system installation)	Cost system operation (electricity)	Potential saving drinking water	Economy monthly average (after system installation)
Unit	m^3	$m^3.mès^{-1}$	$m^3.month^{-1}$	$R\$.month^{-1}$	$R\$.month^{-1}$	$R\$.month^{-1}$	%	$R\$.month^{-1}$
Australian monthly	2946	1333,86	287,00	24215,31	5235,66	665,28	78,48	18314,37
	1457	1333,86	314,00	24215,31	5725,17	665,28	76,46	17824,86
	1000	1333,86	329,00	24215,31	5997,12	665,28	75,33	17552,91
	500	1333,86	361,00	24215,31	6577,28	665,28	72,94	16972,75
	455	1333,86	366,00	24215,31	6667,93	665,28	72,56	16882,10
	250	1333,86	389,00	24215,31	7084,92	665,28	70,84	16465,11
	104	1333,86	411,00	24215,31	7483,78	665,28	69,19	16066,25
	102	1333,86	412,00	24215,31	7501,91	665,28	69,11	16048,12
Monthly simulation	2946	1333,86	283,00	24215,31	5163,14	665,28	78,78	18386,89
	1457	1333,86	307,00	24215,31	5598,26	665,28	76,98	17951,77
	1000	1333,86	322,00	24215,31	5870,21	665,28	75,86	17679,82
	500	1333,86	354,00	24215,31	6450,37	665,28	73,46	17099,66
	455	1333,86	359,00	24215,31	6541,02	665,28	73,09	17009,01
	250	1333,86	382,00	24215,31	6958,01	665,28	71,36	16592,02
	104	1333,86	405,00	24215,31	7375,00	665,28	69,64	16175,03
	102	1333,86	405,00	24215,31	7375,00	665,28	69,64	16175,03

Printed by Books on Demand GmbH, Norderstedt / Germany